Wind Loading and Wind-Induced Structural Response

A state-of-the-art report prepared by the
Committee on Wind Effects of the
Committee on Dynamic Effects of the
Structural Division of the
American Society of Civil Engineers

Published by the
American Society of Civil Engineers
345 East 47th Street
New York, New York 10017-2398

ABSTRACT

This state-of-the-art report presents fundamental and practical aspects of modern wind engineering practice as it relates to structural design. The first chapter includes information on the wind environment. Differences among various types of storms are noted that are of interest in extreme wind climate studies. Elements of engineering micrometeorology included in the chapter consist of descriptions of mean wind profiles for various types of terrain, and of the turbulence structure in the atmosphere. Elements of extreme wind climatology are presented for regions dominated by extratropical storms, hurricane-prone regions, and tornado-prone regions. Information is provided on estimates of extreme winds from short records. The second chapter reviews notions of aerodynamics, structural dynamics, aeroelasticity, and wind tunnel testing. The remainder of the book provides information pertaining to low buildings, tall buildings, slender towers with circular cross-section, hyperbolic cooling towers, and bridges. References are included to numerous primary sources and to pertinent standards and textbooks.

Library of Congress Cataloging-in-Publication Data

American Society of Civil Engineers. Wind Effects Committee.
Wind loading and wind-induced structural response.

Bibliography: p.
1. Wind-pressure. 2. Buildings—Aerodynamics. I. Title.
TA654.5.A53 1987a 624.1'75 87-26642
ISBN 0-87262-625-3

Library of Congress Catalog Card No.: 87-26642
ISBN 0-87262-625-3
Manufactured in the United States of America.

Cover photo: Building destroyed by winds of Cyclone Tracy in Darwin, Australia. Photo by Dr. R.D. Marshall, National Bureau of Standards.

PREFACE

The last ASCE State-of-the-Art report on wind loads and their effects on structures was published in 1961. In the intervening quarter of a century the field of wind engineering has undergone profound changes following a large accumulation of theoretical, computational, and experimental advances. For this reason, the Committee on Dynamic Effects of the Structural Division, ASCE, has appointed a Task Subcommittee charged with the preparation of the present State-of-the-Art Report.

The aim of the Report is twofold. First, it attempts to narrow the gap between practicing engineers and wind engineering specialists by presenting, in simple and readable form, the basic elements from which modern wind engineering practice has evolved. Second, it presents a review of information used for the design of various types of structure.

Chapter 1 of the report includes information concerning the wind environment. Differences among various types of storms are noted, which are of interest particularly in studies of the extreme wind climate. Elements of engineering micrometerology included in this chapter consist of descriptions of the mean wind speed profile for various types of terrain and of the turbulence structure in the atmosphere. The final section covers practical notions of extreme wind climatology in regions dominated by extratropical storms, in hurricane-prone regions, and in tornado-prone regions. This section also includes information on the estimation of extreme winds for regions with short records.

Chapter 2 reviews simple and fundamental notions of aerodynamics, structural dynamics, aeroelasticity, and wind tunnel testing. Familiarity with these notions is necessary for understanding the rationale of modern procedures and techniques used in the estimation of wind loads.

Chapters 3 through 9 provide information pertaining to low buildings, tall buildings, slender towers and stacks, trussed frameworks, guyed towers, hyperbolic cooling towers, and bridges. Owing to space limitations — and to limitations on the resources of the Subcommittee — it was not possible to provide an exhaustive review of these areas. For this reason, references to a large number of primary sources and to various manuals and textbooks have been included.

ACKNOWLEDGEMENTS

The following Subcommittee members contributed to the preparation of the Report:

N. Isyumov, M. J. O'Rourke, J. D. Raggett, D. A. Reed, E. Simiu (Chairman), C. E. Smith, T. Stathopoulos, and J. W. Vellozzi.

The Subcommittee is indebted for valuable comments on various parts of the Report to:

R. F. Abbey, Jr., R. I. Basu, M. E. Batts, J. Biétry, G. R. Cook, T. Galambos, J. D. Holmes, A. Kareem, A. Laneville, H. Liu, K. C. Mehta, D. C. Perry, J. A. Peterka, R. H. Scanlan, D. Surry, and B. J. Vickery.

Partial support for the preparation of the Report was provided by the Center for Building Technology, National Bureau of Standards.

The Subcommittee expresses its appreciation for permission to reproduce the Figures listed below to:

Elsevier Scientific Publishing Company. (Figures 4 and 5, Chapter 3, from the *Journal of Wind Engineering and Industrial Aerodynamics*, Vol. 6, 1980).

Constructional Steel Research and Development Organization (Constrado). (Figs. 6.1.3, 6.2.4, 6.4.2, 6.4.3, 6.4.4, from Constrado publication 1/75, *Wind Forces on Unclad Tubular Structures*.)

Pergamon Press (Figs. 6.11, 6.2.1, 6.2.2, 6.2.3, 6.3.1, 6.3.2, 8.2.4, 8.2.6 and 8.3.1 from *Wind Engineering* (J. E. Cermak, ed.), Proceedings of the Fifth International Conference).

Butterworth Scientific Ltd. (Figure 4.4.1, from *Engineering Structures*, Vol. 5, No. 4, Oct., 1983).

John Wiley and Sons, Inc. (Figure 1.1.1, from E. Simiu and R. H. Scanlan, *Wind Effects on Structures*, Second Edition, 1986).

CONTENTS

1. THE WIND ENVIRONMENT

1.1 BASIC NOTIONS OF METEOROLOGY

Wind, or the motion of air with respect to the surface of the earth, is fundamentally due to differences in the amount of solar heat received by the atmosphere over various areas of the earth's surface. Such differences cause the column of air over these various areas to have different heights. Hence, the respective pressures at points of equal elevation are unequal, and horizontal pressure gradient forces are created, which are responsible for the initiation of the horizontal motion of air. This motion generates additional forces, which include: deflective (Coriolis) forces due to the earth's rotation, centrifugal forces due to the curvature of the path of motion, and retardation forces due to the effect of friction at the earth's surface. In its steady state, the air motion is characterized by a balance among pressure gradient, Coriolis, centrifugal, and friction forces.

Wind forces of concern in structural design are generated both by large scale systems (extratropical cyclones), and by local storms (thunderstorms, foehn winds, winds of the bora type, jet effect winds, tornadoes).

Extratropical Cyclones are produced by the mechanical action of mountain barriers on large scale atmospheric currents, or by the interaction of air masses with relatively uniform properties over horizontal distances comparable to the dimensions of oceans and continents. Extratropical cyclones are observed at latitudes between 35° and 70° and have widths of as much as 1500km. An extratropical cyclone wind speed record is shown in Fig. 1.1.1.

Tropical Cyclones derive their energy from latent heat stored in vaporized ocean water and released as condensation of the water occurs at high altitudes. They originate at latitudes between 5° and 20°, and have diameters of hundreds of kilometers. The central zone of a tropical cyclone, known as the eye, has a diameter of a few tens of kilometers and consists of relatively cloudless and quiescent air. The strongest winds occur near the eye, up to a distance of about 200km from the storm center. Tropical cyclones are referred to as hurricanes if their surface wind speeds exceed 73 mph (32 m/s). Hurricane winds may reach speeds of as much as 200 mph (90 m/s) or so. A satellite view of hurricane Frederick as it crossed the Alabama coast is shown in Fig. 1.1.2.

Thunderstorms are also due to the condensation of water vapor at high altitudes. Raindrops exert viscous drag forces on the air through which they fall and, thus, create a strong downdraft [102].

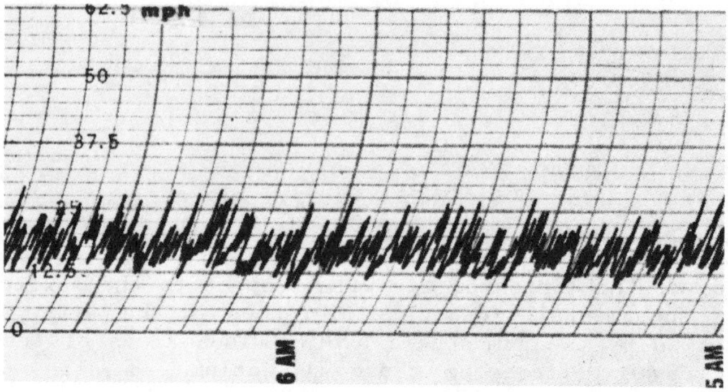

Figure 1.1.1 Wind Speed Record (Extratropical Storm)
From Wind Effects on Structures, Second Edition, by Emil Simiu and Robert H. Scanlan. Copyright 1986 by John Wiley and Sons, Inc. Reprinted by permission of John Wiley and Sons, Inc.

Figure 1.1.2 Satellite View of Hurricane Frederick (NOAA)

Strong winds occur when the downdraft spreads over the ground. This stage in the life-cycle of a thunderstorm usually lasts from 5 to 30 min. and is called the mature stage. A characteristic of thunderstorms is the sharp wind increase, known as first gust, which is associated with the passage of the discontinuity zone between the spreading downdraft and the surrounding air. The occurrence of the first gust is clearly reflected by the thunderstorm wind speed record shown in Fig. 1.1.3.

Foehn Winds occur on the leeward side of mountain ridges. Air forced by the windward mountain slope to rise loses its water vapor content through condensation and precipitation. It then passes over the crest and is forced to descend. In the United States highly turbulent winds of the foehn type occur notably on the slope of the Rocky Mountains where they are known as chinook winds.

Bora type winds occur in areas where a steep slope separates a cold plateau from a warm air zone.

Jet - Effect winds are due to the intensification of winds by topographical configurations producing a convergence of streamlines.

Tornadoes contain the most powerful of all winds. A tornado consists essentially of a vortex of air, typically of the order of 1000 ft (≈300 m) in diameter, which develops within a severe thunderstorm and moves with respect to the ground with speeds of the order of 20-60 mph (≈10-30 m/s). Tornado paths have typically a length of about 10 miles (≈15 km) and are directed in the United States predominantly toward the northeast. Maximum tangential speeds of tornadoes have been estimated to be about 200mph (≈100 m/s), but the possibility that some may actually be considerably higher has not been ruled out.

As in the case of other rotating windstorms, the air motion in a tornado resembles the fluid motion in a stirred cup, where the column of fluid at the center is relatively low. Similarly to the fluid pressure at the bottom of the cup, which is lower at the center than at the edges, the atmospheric pressure at and near the center of a tornado funnel is lower than the normal atmospheric pressure.

Before the passage of a tornado, the pressure in an enclosed structure is equal or close to the normal atmospheric pressure. Upon the occurrence of a tornado strike, powerful explosive forces may be caused by the differences between the pressure within the structure and the lower pressure prevailing within the tornado funnel.

According to Ref. 103, of the windstorm dollar losses incurred by the Factory Mutual System during the period 1962-1973, 34 percent were due to hurricanes and 28 percent to tornadoes. The remaining 38 percent of the windstorm dollar losses were due to other types of storms, including extratropical cyclones and thunderstorms. In terms of number of incidence of windstorm

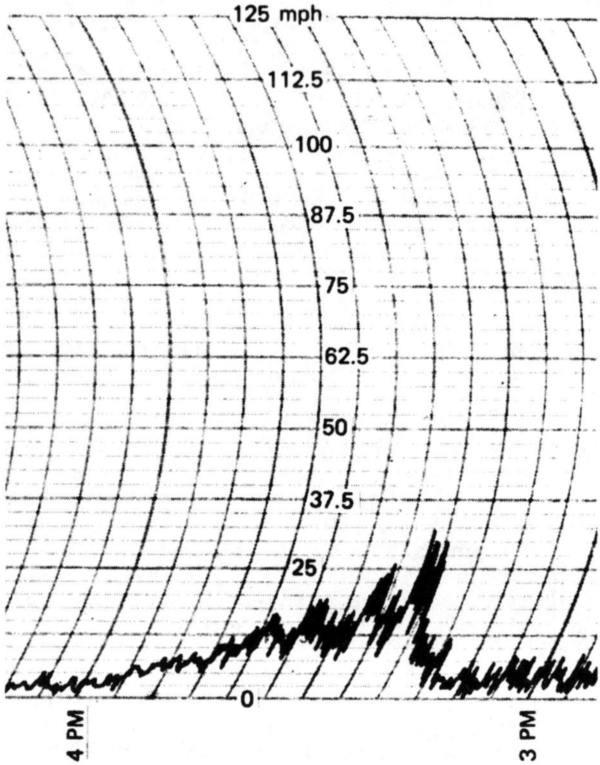

Figure 1.1.3 Thunderstorm Wind Speed Record.

damage, however, hurricanes and tornadoes account for only one of every five wind losses [103].

1.2 ENGINEERING MICROMETEOROLOGY

Topics in micrometeorology of interest to structural engineers include:

- The dependence of mean wind speeds upon height above ground.
- The relation between wind speeds occurring during the same storm at locations with different roughness of terrain.
- The dependence of measured wind speeds upon averaging time.
- The description of features of the atmospheric turbulence, including integral turbulence scales, turbulence intensity, and turbulence spectra and cross-spectra.

The bulk of micrometeorological research conducted to date has been concerned with features of winds in large scale storms. Unless otherwise indicated, the information presented in this section pertains to extratropical storms, or to tropical cyclone winds at relatively large distances from the storm eye.

Most building codes and standards, in the United States and elsewhere, are based on the tacit assumption that information on the micrometeorological features of extratropical storms is applicable to all other types of storm. Although this is not necessarily the case, it is the position of the committee that, in the absence of more information than is currently available, this assumption is acceptable.

In the case of strong winds (i.e., wind speeds exceeding 40 mph [≈20 m/s], say), thorough turbulent mixing tends to produce neutral stratification, just as in a shallow layer of incompressible fluid mixing tends to produce an isothermal state. [Neutral stratification is characterized by the fact that any parcel of air displaced from an initial to a new elevation will be in equilibrium with the surrounding air at that new elevation.] Since the structural engineer is concerned primarily with the effect of strong winds, unless otherwise indicated it will be assumed in this report that the atmospheric stratification is neutral. However, instances where this assumption is not justified do exist, particularly in the case of certain structures experiencing a strong response at relatively low winds speeds (e.g., due to vortex shedding, or aeroelastic instabilities). These instances will be dealt with as needed in the course of this review.

1.2.1 Variation of Mean Wind Speeds With Height Above Ground

1.2.1.1 Mean Wind Speeds Over Terrain With Uniform Roughness

The rougher the terrain, the more effective is the retardation of the mean flow due to the effect of friction at the earth's surface. Also, over terrain of given roughness, the retardation of the mean flow is clearly more effective at lower than at higher elevations. These facts are reflected in the well-known logarithmic law:

$$\frac{u(z)}{u(z_r)} = \frac{\ln \frac{z}{z_o}}{\ln \frac{z_r}{z_o}} \tag{1.2.1}$$

In Eq. 1.2.1, u = mean speed, z = height above ground*, z_r = reference height, and z_o = roughness length. The logarithmic law, regarded by meteorologists as "a cornerstone of micrometeorology" [104] is based upon both theory and experiment. Approximate roughness lengths that may be assumed for structural engineering purposes to characterize various types of terrain are given in Table 1.2.1. Actual roughness lengths for terrains that may be qualitatively classified as belonging to one of the categories I through V may differ somewhat from the values given in Table 1.2.1.

*On account of the finite height of the roughness elements, the following empirical modification of Eq. 1.2.1. is, in principle, required. The quantities z and z_r, rather than denoting heights above ground, are defined as $z = z_g - z_d$ and $z_r = z_{gr} - z_d$, where z_g and z_{gr} denote heights above ground, and z_d is a length known as the zero plane displacement. It is suggested in Refs. 105 and 106 that $z_d \approx H - 2.5\, z_o$, where H = general roof-top level in cities and mean height of trees in forests. However, as shown in Ref. 107, for structural engineering purposes it may be assumed in practice $z_d = 0$ in all cases, so that the empirical modification just mentioned is unnecessary.

Table 1.2.1. Approximate Roughness Lengths for Various Terrain Categories

	Terrain Category*				
z_o	I	II	III	IV	V
	0.005 m (0.0169 ft)	0.007 m (0.230 ft)	0.30 m (0.989 ft)	1.00 m (3.28 ft)	2.50 m (8.2 ft)

---

* Terrain categories are defined as follows: I = open water; II = open terrain; III = suburbs at considerable distance from towns (sites in zones sparsely built-up, with trees, hedges, etc.); IV = towns, densely built-up suburbs, wooded terrain; V = centers of large cities.

The gradient height, δ, is defined as the height above which the retardation of the mean flow by the effects of ground friction becomes negligible. It was shown recently that δ depends upon wind speed and angle of latitude as follows:

$$\delta \simeq 0.25 \frac{u_*}{f} \tag{1.2.2}$$

[108, 109, 110], where the quantity u_* known as friction velocity, is given by the relation

$$u_* = \frac{u(z_r)}{2.5 \ln \frac{z_r}{z_o}} \tag{1.2.3}$$

and f = Coriolis parameter, which varies with latitude as shown, e.g., in Ref. 111, p. 13. The atmospheric layer with depth δ is referred to as the <u>atmospheric boundary layer</u>.

The height, z_ℓ, over which the logarithmic law may be assumed to hold is equal to approximately 0.1 δ [104, 112]. Thus

$$z_\ell \simeq 0.025 \frac{u_*}{f} \tag{1.2.4}$$

Full-scale and laboratory confirmations of Eq. 1.2.4 are illustrated in Refs. 111 (p.40), 113, and 114. It is shown in Ref. 111 that the use of the logarithmic law at heights exceeding z_ℓ is conservative from a structural design viewpoint.

An alternative formulation of the dependence of mean wind speeds upon height above ground is the well-known power law:

$$\frac{u(z)}{u(z_r)} = \left(\frac{z}{z_r}\right)^{\alpha} \qquad (1.2.5)$$

Approximate values of α corresponding to various terrain categories are given in Table 1.2.2.

Table 1.2.2. Approximate Values of Exponent for Various Terrain Categories

Terrain Category*	I	II	III	IV	V
α	0.1	0.14	0.20	0.25	0.30

* see footnote to Table 1.2.1

The power law may be viewed as an approximation to the logarithmic law. The closeness of the approximation depends, for any given pair of values z_o, α, upon the height range being considered. Thus, the profiles given by Eq. 1.2.5 with the values α of Table 1.2.2 are not exact equivalents of the logarithmic profiles (Eq. 1.2.1) with the roughness lengths of table 1.2.1 [115, 116, 117].

The power law is used in both the National building Code of Canada (NBC) [118] and in the American National Standard ANSI A58.1 [119]. However, in NBC, Eq. 1.2.1 is applied to mean hourly speeds while in ANSI A58.1 it is applied to fastest-mile wind speeds (see Section 1.2.2.6). For this reason, the mean speed profiles over open terrain implicit in ANSI A58.1 differ somewhat from those of NBC, even though approximately the same power law exponent ($\alpha \simeq 1/7$) is specified in both documents.

Table 1.2.3. Values of α, and Corresponding Values of z_G, Specified in Ref. 119.

	Terrain			
	Open Water[a]	Open[b]	Suburban[c]	Urban[d]
α	1/10	1/7	1/4.5	1/3
z_G (ft)	700	900	1,200	1,500

Note. 1 ft = 0.3048 m

[a]Exposure D; [b]Exposure C; [c]Exposure B; Exposure A [119].

In ANSI A58.1 it is assumed that Eq. 1.2.5 is applicable up to a height z_G, specified as a function of α as shown in Table 1.2.3.

The height z_G is referred to in engineering practices as the gradient height [119]. This designation, which predates the development of Eq. 1.2.2, is strictly speaking not correct; as previously indicated, the gradient height depends upon wind speed and angle of latitude, rather than having for any given category of terrain the fixed value specified in ANSI A58.1.

1.2.1.2 Relation Between Main Speeds in Different Roughness Regimes

The relation between wind speeds over terrain with a specified roughness length z_o on the one hand, and over nearly open country (e.g., airports) on the other hand can be written as follows:

$$u(z) \simeq u_{ref}\, p \frac{\ln \dfrac{z}{z_0}}{\ln \dfrac{z_{ref}}{(z_0)_{ref}}} \qquad (1.2.6)$$

[107], where $u(z)$ = mean speed at height z over terrain with roughness length z_o, u_{ref} = mean speed at reference height z_{ref} above ground in open terrain, $(z_o)_{ref}$ = roughness length in open terrain. The retardation factor p can be obtained as a function of z_o and $(z_o)_{ref}$ from similarity relations derived by Csanady in 1967 [108]. However, the results so obtained must be subjected to empirical adjustments in the case of built-up terrain (for which $z_o > 0.3$ m or so [107]). Values of p based on full-scale measurements are given in Table 1.2.4.

Table 1.2.4. Values of Factor p for Various Terrain Categories [107]

Terrain Category*	I	II	III	IV	V
p	0.83	1.00	1.15	1.33	1.46

---

* see footnote to Table 1.2.1

Numerical Example. Calculate the mean wind speed u(z) at 100ft (30.5 m) elevation over category IV terrain (i.e., densely built-up suburbs), if the mean speed at 32.8 ft (10 m) above ground at an airport site (open terrain) is u_{ref} = 40 mph (22.25 m/s).

Using Eq. 1.2.6 with values taken from Tables 1.2.1 and 1.2.4

$$u(100) = 50 \times 1.33 \frac{\ln \frac{100}{3.28}}{\ln \frac{32.8}{0.23}}$$

A relation similar to Eq. 1.2.6 can be written in terms of the power law formulation:

$$u(z) = u_{ref}\, b \left(\frac{z}{z_{ref}}\right)^{\alpha} \qquad (1.2.7)$$

This relation was proposed in 1961 by Davenport [120] in conjunction with the relation

$$b = \left(\frac{z_{Gref}}{z_{ref}}\right)^{\alpha ref} \left(\frac{z_{ref}}{z_G}\right)^{\alpha} \qquad (1.2.8)$$

in which the subscript "ref" indicates that z_G and α correspond to open terrain conditions.

<u>Numerical Example</u>. Calculate the mean wind speed u(z) at 100 ft (30.5 m) elevation over suburban terrain if the mean speed at 32.8 ft (10 m) above ground at an airport site (open terrain) is u_{ref} = 50 mph (22.25 m/s).

Using Eqs. 1.2.7 and 1.2.8 with the values of ANSI A58.1 Table 1.2.3):

$$u(100) = 50\left(\frac{900}{32.8}\right)^{1/7} \left(\frac{32.8}{1200}\right)^{1/4.5} \left(\frac{100}{32.8}\right)^{1/4.5} = 46.2 \text{ mph } (20.7 \text{ m/s})$$

If the values of NBC are used: u(100) = 43.45 mph (19.47 m/s).

The values obtained by Eqs. 1.2.6 and 1.2.7 are, in this example, relatively close. However, in other cases differences of the order of 10 percent or even more may arise, particularly in the case of urban terrain (centers of large cities) where recent research [107, 115, 116, 117] indicates that the speed reductions over urban terrain inherent in Refs. 118 and 119 appear to be too strong.

1.2.1.3 <u>Mean Wind Speeds Downwind of a Roughness Change</u>

If the wind flows from a less rough to a rougher zone (e.g., from open water onto the shore, or from open terrain into a town), the rougher terrain will retard the flow more effectively than the terrain upwind of the roughness change. The region within which this increased retardation effect occurs is referred to as the internal boundary layer. The thickness of this layer grows with distance from the roughness change. However, past a sufficiently large distance (about 5 km) this growth becomes negligibly small and the flow regime corresponding to the higher roughness may be considered to be established.

It is shown in Ref. 107 that an acceptable approximation for the height of the internal boundary layer, δ_i, is

$$\delta_i(x) \simeq \frac{x}{12.5} \qquad (1.2.9)$$

where x is distance downwind from roughness change. Above elevation δ_i the profile is the same as upwind from the roughness change. Below elevation δ_i the profile is assumed to be logarithmic. The profiles above and below elevation δ_i are continuous at the elevation (Fig. 1.2.1).

Numerical Example. Determine the mean wind profile at a distance x = 1310 ft (400 m) downwind of a roughness change from z_o = 0.23 ft (0.07 m) [open terrain] to z_o = 3.28 ft (1.00 m) [densely built-up suburbs] - see Table 1.2.1.

The mean wind profile above an elevation δ_i = 1310/12.5 = 105 ft (32 m) (Eq. 1.2.9) is the same as over the terrain upwind of the roughness change, i.e., using Eq. 1.2.1,

$$u(z) = u_{ref} \frac{\ln \frac{z}{0.23}}{\ln \frac{32.8}{0.23}}$$

$$= 0.202\, u_{ref} \ln \frac{z}{0.23} \qquad (z \geqslant 105 \text{ ft})$$

where u_{ref} = mean speed at elevation z = 32.8 ft (10 m) over the open terrain (z in feet). At z = 105 ft (32 m), it follows from the preceding relation that

$$u(105) = 0.9997\, u_{ref}$$

For z < 105 ft (32 m)

$$u(z) = u(105) \frac{\ln \frac{z}{3.28}}{\ln \frac{105}{3.28}}$$

$$= 0.287\, u_{ref} \ln \frac{z}{3.28}$$

(z in feet).

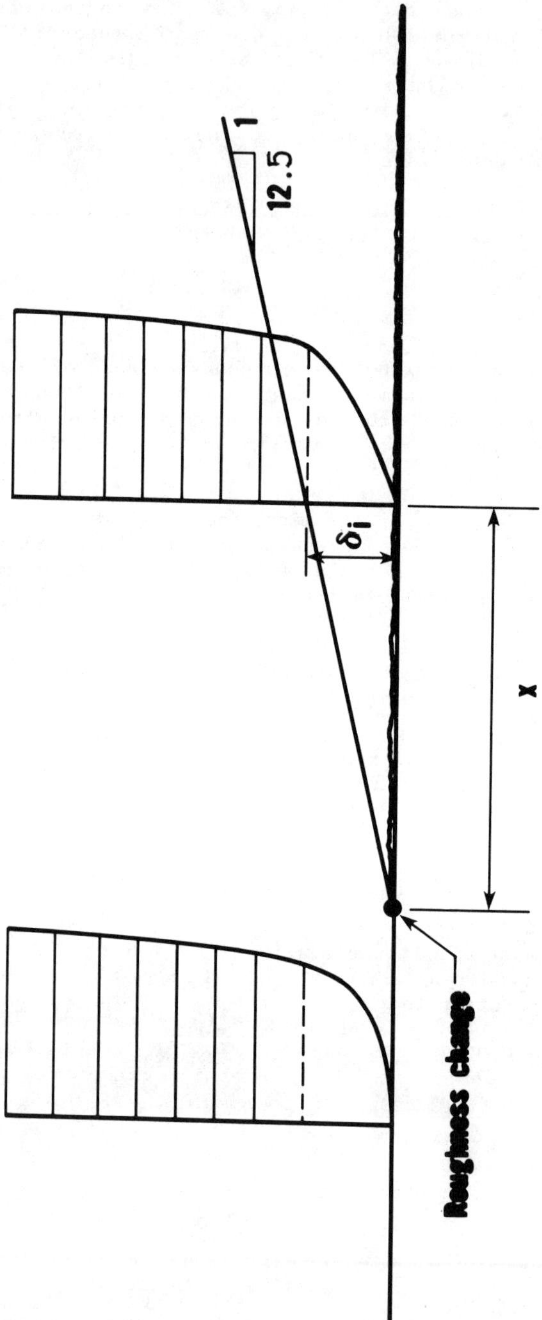

Figure 1.2.1 Mean Wind Profiles Upwind and Downwind of a Roughness Change

The above procedure can similarly be applied downwind of two successive changes in terrain roughness [107].

1.2.1.4 Mean Winds Speeds Over Escarpments and Hills

Theoretical, full scale, and wind tunnel studies pertaining to the amplification of wind speeds over two-dimensional escarpments and hills are reported in Ref. 121 to 129 and are summarized in Ref. 111.

1.2.1.5 Mean Wind Speeds in Hurricanes and Local Storms

It has been pointed out in Ref. 130 that mean wind profiles may differ in tropical cyclones from those typical of extratropical storms. One reason for such differences is that in the region of highest tropical cyclone winds the flow paths are relatively strongly curved. Associated with these large curvatures are centrifugal forces which may affect the shape of the wind profile. A numerical study reported in Ref. 131 concluded that the influence of these forces upon the wind profile in the lowest 400 m of the atmosphere is negligible, so that the logarithmic law (Eq. 1.2.1) would be also applicable to tropical cyclone winds. However, owing to the lack of reliable experimental data on tropical cyclone wind profiles, this conclusion can be regarded as only tentative.

In the case of thunderstorms, the notion of gradient height has no meaning since strong winds are caused by the spreading of a downdraft on the surface of the ground. Measurements reported in Ref. 132 suggest that, up to about 300 ft (91 m) above ground, thunderstorm wind speeds vary with height in accordance with the logarithmic law. Above this elevation it appears that the variation of wind speeds with height is generally small.

1.2.2. Atmospheric Turbulence

An examination of Fig. 1.1.1 shows that the recorded wind speeds vary randomly with time. This variation is due to the turbulence of the wind flow.

Information on the main features of the atmospheric turbulence is useful from a structural engineering viewpoint for the following reasons. First, flexible structures may exhibit resonant amplification effects induced by the velocity fluctuations. Second, rigid structures and structural parts are subjected to time-dependent wind loads with fluctuations due, at least in part, to atmospheric turbulence. Third, aerodynamic test results obtained in the laboratory may be influenced to a considerable extent by the features of the turbulence in the wind flow. Fourth, owing to the unsteadiness of the wind flow, measured wind speeds depend upon the averaging time. For example, if the averaging time is 1 hour, say, the mean wind speed in Fig. 1.1.1 is about 19 mph (8.5 m/s), whereas the peak speed (which may be assumed to be averaged over a time interval of the order of 1 sec, i.e., the response time of the anemometer)is about 29 mph (12.9 m/s). Most principal weather stations in the United States use equipment that automatically records wind speeds averaged over an interval equal to the time required for the passage over the anemometer of a volume of air with a horizontal length of one mile. The highest wind speed recorded by this equipment during any one day, month, or year is known as the fastest-mile wind speed for that day, month, or year. From the definition of the fastest-mile wind speed, it follows that the averaging time, t, in seconds, for a recorded fastest-mile wind speed, v_f, in miles per hour, is

$$t = 3600/v_f \tag{1.2.10}$$

Clearly, because the atmospheric flow is turbulent, the significance to the structural engineer of any given wind speed is inseparable from the averaging time associated with its measurement. For example, wind speeds specified in the National Building Code of Canada [118], which are given in terms of hourly means, do not have the same meaning as the wind speeds specified by the American National Standard A58.1 [119], which are given in terms of fastest-miles. A comparison between wind speeds specified in Refs. 118 and 119 is presented for illustration purposes in Section 1.2.2.6.

1.2.2.1 Integral Scales of Turbulence

Integral scales of turbulence are measures of the average size of the turbulent eddies of the flow. To simulate correctly the wind loads acting on any given area, the ratio between integral scales in the laboratory and in the atmosphere must be equal to the geometric scale, i.e., to the ratio between the dimensions of the model and the corresponding dimensions of the prototype. Strictly speaking, a model of any given size will thus be representative of a prototype with dimensions equal to those of the model multiplied by the ratio between the actual integral scales of the atmospheric turbulence and the corresponding integral scales of the turbulence created in the wind tunnel. The fact that the geometric scale cannot be chosen arbitrarily but is imposed by that ratio explains the need for information on the integral scales of the atmospheric turbulence.

There are altogether nine integral scales of turbulence, corresponding to the three dimensions of the longitudinal, transverse, and vertical turbulence eddies. (For a mathematical definition of the integral scales of turbulence, see Refs. 111 and 133). For example, L_u^x, L_u^y, L_u^z are, respectively, measures of the longitudinal, transverse, and vertical size of the eddies associated with the longitudinal velocity fluctuations. Values of L_u^y and L_u^z that are small compared to the dimensions of a panel normal to the mean wind indicate that the effect of the longitudinal velocity fluctuations upon the overall wind loading acting on that panel is small. However, if L_u^y and L_u^z are large, the eddy will envelop the entire panel, and that effect will be significant.

It appears that L_u^x is influenced by unpredictable mesometeorological phenomena and varies widely from experiment to experiment. For example, in open terrain, measured integral scales Lu^x were found to vary between 120 m and 630 m at 11 m (average value: 350 m); and between 60 m and 460 m at 31 m elevation (average value: 200 m) [134]. Tentative models for Lu^x, Lu^y, Lu^z are summarized in Refs. 111, 115, 135 and 136.

1.2.2.2 Turbulence Intensity

The turbulence intensity is defined as the ratio between the r.m.s. of the velocity fluctuations and the mean wind speed. For the longitudinal velocity fluctuations

$$I(z) = \frac{\sqrt{\overline{u'^2(z)}}}{u(z)} \qquad (1.2.11)$$

where u' (z) = longitudinal velocity fluctuations about the mean wind speed u(z).

The mean square value of the longitudinal turbulence fluctuations can be expressed empirically as follows

$$\overline{u'^2} = \beta u_*^2 \qquad (1.2.12)$$

where u_* is given by Eq. 1.2.3. Values of β suggested in Ref. 107 for structural design purposes are given in Table 1.2.5.

Table 1.2.5. Values β for Various Terrain Categories

Terrain Category*	I	II	III	IV	V
β	6.00	6.00	5.25	4.85	4.00

* See footnote to Table 1.2.1

It has been noted in Ref. 116 that actual values of β may be somewhat lower than 4.00 in Category V terrain, and higher than 6.00 in Category I terrain [117].

From Eqs. 1.2.11, 1.2.12, 1.2.1, and 1.2.3, it follows that

$$I(z) = \frac{\beta^{1/2}}{2.5 \ln \frac{z}{z_0}} \qquad (1.2.13)$$

1.2.2.3 Turbulence Spectra

A randomly fluctuating phenomenon such as that depicted in Fig. 1.1.1 can be conceived of as the superposition of a large number of harmonic fluctuations with frequencies ranging from zero to infinity. The spectral representation of turbulence is related to this concept, and provides information on the contributions of fluctuating components with various frequencies to the mean square value of the turbulent velocity fluctuations at a point, or on the frequency content of the velocity fluctuations at the point. This information is primarily of interest in situations where the turbulence induces fluctuating forces that cause significant resonant amplification effects on flexible structures. From the definition of spectra, the following relation holds:

$$\overline{u'^2(M)} = \int_0^\infty S_{u'}(M, n)\, dn \qquad (1.2.14)$$

where $u'(M)$ = velocity fluctuations at point M, $S_{u'}$ = spectrum of velocity fluctuations, and n = frequency (see e.g., Ref. 111). The ordinates of the spectrum, i.e., the contributions of components with various frequencies, decrease as the frequencies increase. For this reason the effect of turbulence fluctuations upon the resonant response of rigid buildings (i.e., buildings with high natural frequencies)is generally small. Also, in the frequency range of interest in the structural design of buildings, the ordinates of the longitudinal velocity spectra decrease as the height above ground increases [137, 138].

For a mathematical definition of turbulence spectra the reader is referred to [133] or [111]. Various expressions proposed in the literature for the spectra of longitudinal, lateral and vertical turbulent velocity fluctuations are given in Refs. 111 and 115, and are discussed in Refs. 137, 138 and 139.

1.2.2.4 Cross - Spectra of Longitudinal Velocity Fluctuations

As previously mentioned, determining the resonant response to turbulence-induced pressures requires information on the frequency content of the velocity fluctuations at any given point. In addition, information is needed on the degree to which harmonic components of velocity fluctuations occurring at any two points are mutually coherent. This information is expressed in terms of cross-spectra of the velocity fluctuations at those points (see, e.g., Ref. 111).

It is intuitively clear that the extent to which harmonic components of velocity fluctuations occurring at any two points are mutually coherent decreases as the separation distance increases. Moreover, this decrease is faster for higher frequencies. The following empirical expression was proposed to account for this behavior:

$$S^{C}_{u'_{M_1} u'_{M_2}}(n) \simeq S_{u'}^{1/2}(M_1, n)\, S_{u'}^{1/2}(M_2, n)\, e^{-\bar{f}} \tag{1.2.15a}$$

[140] where $S^{C}_{u'_{M_1} u'_{M_2}}$ = real part of cross-spectrum of longitudinal

velocity fluctuations at points M_1 and M_2, and

$$\bar{f} = \frac{n[C_{1z}(z_1 - z_2)^2 + C_{1y}(y_1 - y_2)^2]^{1/2}}{u(10)} \tag{1.2.15b}$$

In Eqs. 1.2.15, y_1, y_2, and z_1, z_2 are the coordinates of points M_1, M_2, respectively, the line M_1M_2 is normal to the direction of the mean wind, u(10) = mean speed at 10 m above ground, and the empirical exponential decay coefficients are determined experimentally and may be assumed in applications to have, approximately, the values $C_{1z} \simeq 6$, $C_{1y} \simeq 9$ [140]. An alternative form of $\bar{f}$ is

$$\bar{f} = \frac{n[C_z(z_1 - z_2)^2 + C_y(y_1 - y_2)^2]^{1/2}}{\frac{1}{2}[u(z_1) + u(z_2)]} \tag{1.2.16}$$

On the basis of wind tunnel experiments, Vickery proposed for strong winds the values $C_z = 10$, $C_y = 16$ [140].

It appears that the exponential decay coefficients are in reality functions of wind speed, terrain roughness, and height above ground. However, in spite of various recent theoretical and experimental investigations [142, 143, 144, 145, 146, 147], it has not been possible to date to improve significantly upon the model proposed in Refs. 140 and 141.

1.2.2.5 Turbulence in Flow Over Hills

According to full-scale measurements reported in Refs. 122 and 123, the effect of a hill upon the turbulence intensity and the turbulence scales is negligible for practical purposes, except in the wake of the hill crest if the slope is large (say, 15 percent or more). Wind tunnel measurements of the turbulence intensities over a steep bell-shaped hill are reported in Ref. 120. According to these measurements, the turbulence intensity near the ground is higher by almost 100 percent in the vicinity of the upwind foot of the hill than would be the case in the absence of the hill.

1.2.2.6 Dependence of Wind Speeds on Averaging Time

This dependence may be expressed, approximately, as

$$u_t(z) \simeq u(z) + C(t)\,\overline{u'^2}^{1/2} \qquad (1.2.17)$$

where $u_t(z)$ = most probable value of wind speed averaged over t seconds, $u(z)$ and $\overline{u'^2}$ are given by Eqs. 1.2.1 and 1.2.9 in which the parameter values of Tables 1.2.1 and 1.2.5 may be used, and $C(t)$ has the values of Fig. 1.2.2, which is based on Ref. 148.

Numerical Example. Calculate the mean hourly speed 32.8 ft (10 m) above ground in open terrain, u (32.8), corresponding to a recorded fastest-mile wind speed v_f = 72 mph (32 m/s) at 32.8 ft (10 m) above ground in open terrain.

From Eq. 1.2.10, t = 3600/72 = 50 sec, i.e., C(t) = 1.35 (Fig. 1.2.2). From Eqs. 1.2.17, 1.2.12, 1.2.3, and Tables 1.2.1 and 1.2.5 (Category II terrain),

$$72 = u(32.8) + 1.35\,\sqrt{6}\,\frac{u(32.8)}{2.5\,\ln\dfrac{32.8}{0.23}}$$

i.e., u(32.8) ≅ 57 mph (25 m/s).

The preceding result illustrates typical differences between the numerical values of the wind speeds specified, for regions adjacent to the U.S.-Canadian border, by the National Building Code of Canada [118] in terms of mean hourly winds on the one hand, and by the American National Standard ANSI A58.1 [119] in terms of fastest-miles on the other hand.

1.3 EXTREME WIND CLIMATOLOGY

Extreme wind climatology may be defined as a set of probabilistic statements on future extreme wind conditions at various geographical locations.

Techniques used in extreme wind climatology differ according to whether, at the location of concern, winds of the type that is of interest from a structural design viewpoint occur every year or at larger intervals. In the first case (referred to as the case in which the wind climate is well-behaved), the set of data consisting of the maximum wind speed recorded every year constitutes the data base from which statistical inferences on the extreme wind can be made. In the second case (e.g., in hurricane prone regions, where hurricanes are the type of wind controlling the design), wind speed data recorded in years where hurricanes do not occur are generally of secondary, if any, interest to structural engineers. Therefore, a data base consisting of the maximum annual winds might well

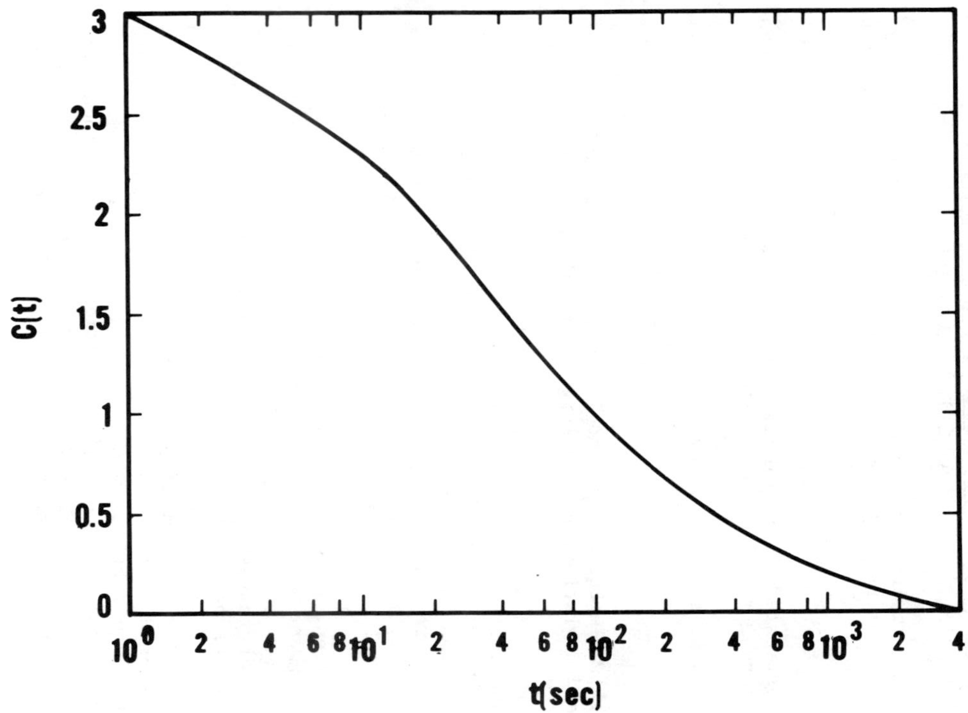

Figure 1.2.2 Dependence of coefficient C(t) on Time, t.

provide a misleading picture of the extreme climate at the location being investigated [111].

1.3.1 Extreme Wind Speeds in Well-Behaved Climates

1.3.1.1 Estimation of Extreme Wind Speeds With Specified Probability of Occurrence in Any One Year

Statistical studies have shown that, for structural design purposes, it is reasonable to assume the validity of the expression

$$u_N = a + b \ln N \quad (1.3.1)$$

where u_N = estimated value of extreme wind speed with a mean recurrence interval N(in years),

$$a = \bar{X} - 0.45\ s \quad (1.3.2)$$

$$b = 0.78\ s \quad (1.3.3)$$

[149, 150, 151]. In Eqs. 1.3.2 and 1.3.3, $\bar{X}$ = sample mean, and s = sample standard deviation, of the largest annual speeds.

The largest annual wind speed data must form a micrometeorologically homogeneous set, i.e., all data must have been measured over terrain with the same roughness and at the same height above ground, and averaged over the same time interval. If these conditions are not satisfied, the data must be adjusted to a common terrain roughness, height above ground, and averaging time, by using Eqs. 1.2.1, 1.2.6, and 1.2.17.

It can be seen from Eq. 1.2.10 that to two different fastest-mile speeds, v_{f1} and v_{f2}, there correspond two different averaging times, t_1 and t_2. Nevertheless, if all the data in a sample are given in terms of fastest-mile speeds, such differences are small and may in practice be ignored.

One potential source of large errors in the estimation of u_N is the conversion to speeds over open terrain of data recorded over, or downwind of, built-up terrain. To avoid such errors it has been proposed that in certain situations high altitude wind speed data - which are not influenced by ground friction and, hence, by the terrain roughness - be used in lieu of wind speed data recorded near the ground [152].

Numerical Example. The following largest annual wind speeds (fastest-miles) were recorded at one station over 16 consecutive years at 32.8 ft (10 m) above ground in open terrain:

56, 62, 69, 65, 63, 51, 51, 50, 61, 55, 57, 56, 59, 47, 47, 51

Estimate the wind speed with a mean recurrence interval N = 50 years (fastest-mile) at 32.8 ft (10 m) above ground in open terrain. For this data sample mean and standard deviation are $\bar{X}$ = 56.46 mph and s = 6.57 mph, respectively. From Eqs. 1.3.1, 1.3.2, and 1.3.3, the estimated wind speed being sought is

$$u_{50} = 73.55 \text{ mph}$$

In statistical terminology, Eqs. 1.3.1, 1.3.2 and 1.3.3 constitute an estimation procedure, based on the method of moments, of a percentage point function described by the Gumbel (or Extreme Value Type I) distribution. As shown in Ref. 151, more precise estimation procedures for the Gumbel distribution are available and may be used. However, since the differences in precision between the method of moments and the other estimation methods is small, Eq. 1.3.1 can be regarded as adequate for most structural engineering design applications [151].

Probabilistic models that in certain cases fit the largest annual wind speeds more closely than the Gumbel distribution have been proposed in the literature [153, 154, 155, 156]. Generally, the choice of any of these models does not affect very significantly the estimated extreme winds corresponding to mean recurrence intervals $N < 50$ years. However, according to Ref. 156, over large portions of the U.S. territory the use of the Gumbel distribution appears to lead to overestimates of the extreme wind speeds with very long mean recurrence intervals. This result is of interest in structural reliability calculations.

An attempt to model extreme wind speeds in which thunderstorm wind statistics are used has been reported in Ref. 157.

It has been shown in Refs. 158 and 159 that Eqs. 1.3.1, 1.3.2, and 1.3.3 can be used to estimate the extreme wind speeds with mean recurrence interval N (in <u>months</u>), provided that $\bar{X}$ and s are defined, respectively, as the sample mean and the sample standards deviation of the largest <u>monthly</u> wind speeds. A sample of 36 consecutive largest monthly speeds is usually sufficient to obtain respectable estimates of the extreme wind speed with a mean recurrence interval N = 600 months (50 years) - see Ref. 111 for details.

1.3.1.2 <u>Sampling Errors in Estimation of Extreme Wind Speeds</u>

The extent to which estimates obtained by using Eq. 1.3.1 are reliable depends upon the size of the data sample on which the estimates are based. It is clear that the estimates will improve as the sample size increases and vice-versa.

Using statistical theory it can be shown that the standard deviation of the sampling errors in the estimation of extreme wind speeds is given, approximately, by the expression

$$\sigma(u_N) \simeq 0.82\,(\ell n^2 N + 0.17\,\ell n N + 1.06)^{1/2}\,\frac{s}{\sqrt{n}} \qquad (1.3.4)$$

where n = sample size and s and N are defined as in Eqs. 1.3.1 and 1.3.2 [151]. Under certain simplifying assumptions it can be stated, with approximately 68 percent, 95 percent, and 99 percent, confidence, that the true value of u_N is contained within the interval $u_N \pm \sigma(u_N)$, $u_N \pm 2\sigma(u_N)$, and $u_N \pm 3\sigma(u_N)$, respectively.

Numerical Example. Given the data of the numerical example in Section 1.3.1.1, calculate the standard deviation of the sampling error in the estimation of u_{50}

From Eqs. 1.3.4 and 1.3.5, in which N = 50, n = 16, and s = 6.57 mph, $\sigma(73.55) \simeq 5.55$ mph. It can be stated with 95 percent confidence that the true value of u_{50} is contained in the interval 73.55 ± 2 x 5.55, i.e., the interval 62.45 mph-84.65 mph. Note that had the estimate of u_{50} been based on a 32 year, rather than 16 year, record, the result $\sigma(u_N) = 5.55/\sqrt{2} = 3.92$ would have been obtained.

1.3.1.3 Dependence of Extreme Wind Speeds Upon Direction

In a large number of applications the aerodynamic loads depend strongly upon the direction of the wind velocity. This is true, for example, of wind pressures on cladding, for which aerodynamic tests are routinely conducted for various orientations of the structure with respect to wind direction.

In current procedures for assessing cladding loads, direction dependent aerodynamic information has until recently been commonly used in conjunction with direction-independent climatological information. This practice tends to result in large overestimates of wind loads corresponding to any given probability of occurrence [160]. To show why this is the case the process whereby direction-dependent aerodynamics is combined with direction-independent extreme wind climatology is illustrated schematically below.

Assume that the extreme wind speeds recorded in two consecutive years for the four directions of the compass are the following:

	E	N	W	S
Year 1	10	50	30	20
Year 2	15	30	60	18

Assume further that the aerodynamic pressure coefficients for the cladding panel of concern corresponding to winds blowing from these directions are

E	N	W	S
1.00	2.00	1.00	2.00

The procedure of Ref. 119 consists in effect of assuming that the maximum pressures on the panel occurring in years 1 and 2 are proportional to 2.00×50^2 and 2.00×60^2, respectively. Actually, while the maximum pressure in year 1 would indeed be proportional to 2.00 x 50^2, the maximum pressure in year 2 would be proportional to the largest of the products 1.00 x 15^2, 2.00 x 30^2, 1.00 x 60^2, and 2.00 x 18^2, i.e., to 1.00 x 60^2. Thus, while ignoring wind directionality effects would result in a maximum pressure in years 1 and 2 proportional to 2.00 x 60^2, a procedure that would take into account the dependence of the extreme winds upon the directionality of the wind climate would estimate the pressure to be proportional to only 2.00 x 50^2.

For additional details, the reader is referred to Ref. 111, which also contains an assessment of procedures for estimating wind directionality effects proposed in Refs. 161, 162, and 163.

1.3.2 Extreme Wind Speeds in Hurricane-Prone Regions

The number of hurricane wind speed data recorded at any one station over a period of, say, 30 years is usually so small that the sampling errors (Section 1.3.1.2) inherent in estimates of the extreme wind speeds from those data would be unacceptably large. Primarily for this reason, a special procedure for estimating hurricane wind speeds has been developed in Ref. 164. The procedure makes use of statistics on climatological characteristics of hurricanes, including atmospheric pressures and radii of maximum wind speeds within hurricanes, storm translation speeds, and rate of occurrence of hurricanes in any given region. These characteristics can be related to the hurricane wind speeds at the site being investigated and to their probabilities of occurrence [165, 166]. Various versions of the procedure developed in Ref. 164 have been applied in Ref. 167, 168, 169 and 170. Calculated sampling errors in the estimation of hurricane wind speeds are reported in Ref. 171, where it is shown that the standard deviation of these errors is typically 5 percent to 10 percent of the estimated wind speed.

Analyses pertaining to the effect of observation and modeling errors upon the estimation of hurricane wind are reported in Ref. 170. Estimates of the dependence of hurricane wind speeds upon direction are given in Ref. 173, and in greater detail (on magnetic tape) in Ref. 174.

1.3.3 Estimates of Extreme Wind Speeds in the United States

Estimated fastest-mile wind speeds with mean recurrence intervals N = 50 years are published in Refs. 172 and 175, where detailed information on extreme wind speeds corresponding to various probabilities of occurrence is available.

In the United States, the number of weather stations with records that can be used for the estimation of extreme wind speeds is of the order of one hundred. For this reason, records taken at one station are used to make inferences on the extreme wind climate over areas of tens of thousands of kilometers. Whether this is appropriate or whether a wind climate microzonation is required is established by examining the basic atmospheric flow patterns in the area under consideration, wind data at secondary weather stations, and historical damage experience [186].

Data and information on extreme directional wind speeds are given in Refs. 176 and 177.

1.3.4 Climatology of Tornadoes

Let A_0 be some large reference area, and n the estimated average number of tornado occurrences per year in the area A_0. The estimated probability that a tornado will strike a particular location in any one year is assumed to be

$$P(T) = (T) = n \frac{\bar{a}}{A_0} \tag{1.3.6}$$

where $\bar{a}$ = estimated mean area of individual tornado path.

The probability $P(T,V_0)$ that a tornado with maximum wind speeds higher than some specified value V_0 will strike a location in any one year can be written as

$$P(T,V_0) = P(V_0)P(T) \tag{1.3.7}$$

where $P(V_0)$ = probability that the maximum speed is higher than V_0, given that a tornado has occurred. The estimation of $P(T,V_0)$ relies on relatively few and uncertain data -- inferred mostly from observations of damage -- on tornado occurrences, path areas, and wind speeds, and largely on subjective extrapolations from these data to small probability levels. According to estimates presented in Ref. 178, maximum tornado wind speeds corresponding to a probability $P(H,V_0) \simeq 10^{-7}$ in any one year vary between 400 mph in Oklahoma and Nebraska to 240 mph in Northern California and Oregon. Alternative estimates presented in Ref. 181 suggest that these speeds may be reduced significantly. Nuclear power plant design data on tornado wind speeds, pressure drops, and pressure drop rates in the United States are summarized in Refs. 178 and 111. Information (including pertinent references) on estimated tornado-borne missile speeds is summarized in Refs. 111 and 185. Useful engineering contributions to the topic of tornado climatology are presented in Refs. 179 through 184.

REFERENCES

101. Wind Forces on Structures, Final Report on the Task Committee on Forces of the Committee on Loads and Stresses of the Structural Division, Transactions, ASCE, Vol. 126, Part II, 1961, pp. 1124-1198.

102. Thunderstorm, Report of the Thunderstorm Project, U.S. Department of Commerce, Washington, DC, 1949.

103. Wind Forces on Buildings and Other Structures, Loss Prevention Data, 1-7, Factory Mutual System Engineering Corp., June 1974.

104. Tennekes, H., "The Logarithmic Wind Profile," Journal of Atmospheric Sciences, Vol. 30, July 1972, pp. 234-238.

105. N. C. Helliwell, "Wind Over London" in Proceedings of the Third Internationsl Conference on Wind Effects on Buildings and Structures, Tokyo, 1971, Saikon, 1972, pp.

106. Oliver, H. R., "Wind Profiles in and Above a Forest Canopy," J. Royal Meteorol. soc. Vol. 97, 1971, pp. 548-553.

107. Biétry, J., Sacré, C., Simiu, E., "Mean Wind Profiles and Changes of Terrain Roughness," Journal of the Structural Division, ASCE, Vol. 104, October 1978, pp. 1585-1593.

108. Csanady, G. T., "On the Resistance Law of a Turbulent Ekman Layer," Journal of Atmospheric Sciences, Vol. 24, September 1967, pp. 467-471.

109. Tennekes, H.and Lumley, J. L., A First Course in Turbulence, MIT Press, Cambridge, MA.

110. Monin, A. S., and Yaglom, A. M., Statistical Fluid Mechancs of Turbulence, Vol. 1, MIT Press, Cambridge, MA, 1971.

111. Simiu, E., and Scanlan, R. H., Wind Effects on Structures, Second Edition, John Wiley & Sons, New York, NY., 1986.

112. Monin, A. S., and Obuchow, A. M., "Fundamentale Gesetzmaessigkeiten der turbulenten Vermischung in der bodennahen Schicht der Atmosphaere," in Sammelband zur statistischen Theorie der Turbulenz, H. Goering, ed., Akademie-Verlag, Berlin, 1958, pp. 199-226.

113. Thuillier, R. H., and Lappe, V. O., "Wind and Temperature Profile Characteristics from Observations on a 1400 ft Tower," Journal of Applied Meteorology, Vol. 3, June 1964, pp. 299-306.

114. Howroyd, G. C., and Slawson, P. R., "The Characteristics of a Laboratory Produced Turbulent Ekman Layer," Boundary-Layer Meteorology, Vol. 8, 1975, pp. 201-219.

115. Counihan, J., "Adiabatic Atmospheric Boundary Layers: Atmospheric Environment, Vol. 9, No. 10, October 1975, pp. 871-905.

116. Counihan, J., Discussion of "Mean Wind Profiles and Change of Terrain Roughness" by Biétry, J., Sacré, C., and Simiu, E., Journal of the Structural Division, ASCE, June 1978, p. 1225.

117. Biétry, J., Sacré, C., and Simiu, E., "Closure to Mean Wind Profiles and Change of Terrain Roughness," Journal of the Structural Division, ASCE, 1980.

118. Canadian Structural Design Manual, Supplement No. 4 tothe National Building Code of Canada. Associate Committee on the National Building Code and National Research Council of Canada, Ottawa, Canada 1975.

119. American National Standard Building Code Requirements for Minimum Design Loads in Buildings and Other Structures. A58.1, American National Standards Institute, New York, NY, 1982.

120. Davenport, A. G., "The Spectrum of Horizontal Gustliness Near the Ground in High Winds," J. Royal Meteorol. Soc., 87, 1961, pp. 194-211.

121. Jackson, P. S., The Influence of Local Terrain Features on the Site Selection for Wind Energy Generating System, BLWT-1-1979, The University of Western Ontario, London, Canada, 1979.

122. Sacré, C., "An Experimental Study of the Airflow Over a Hill in the Atmospheric Boundary Layer," Boundary Layer Meteorology, Vol. 17, 1979, pp. 381-401.

123. Bradley, E. F., "An Experimental Study of the Profiles of Wind Speed, Shearing Stress and Turbulence at the Crest of a Large Hill," J. Royal Meteorol. Soc., 106, 1980, pp. 181-123.

124. Bowen, A. J., and Lindley D., "A Wind Tunnel Investigation of the Wind Speed and Turbulence Characteristics Close to the Ground Over Various Escarpment Shapes," Boundary-Layer Meteorology, Vol. 12, 1977, pp. 259-271.

125. Meroney, R. N., Sandborn, V. A., Bowmeester, R., and Rider, M., "Wind Tunnel Simulation of the Influence of Two-Dimensional Ridges on Wind Speed and Turbulence," Proceedings, International Symposium on Wind Energy Systems, Cambridge, U.K., 1976.

126. Sacré, C., Structure des vents forts au passage d'une discontinuité de rugosité mer-terre, EN-CL179.4L, Centre Scientifique et Technique du Bâtiment, Nantes, France, 1979.

127. Jackson, P. S. and Hunt, J. C. R., "Turbulent Wind Flow Over a Low Hill," J. Royal Met. Soc., Vol. 101, 1975, pp. 929-955.

128. Taylor, P., "Some Numerical Studies of Surface Boundary-Layer Flow Above Gentle Topography," Boundary-Layer Meteorology, Vol. 11, 1977, pp. 439-465.

129. Hunt, J. C. R. and Snyder, W. H., "Experiments on Stable and Neutrally Stratified Flow Over a Model Three-Dimensional Hill," Journal of Fluid Mechanics, 1980, Vol. 96, part 4, pp. 671-704.

130. Smith, R. K., "The Surface Boundary Layer of a Hurricane," Tellus, 20, 1968, pp. 473-484.

131. Simiu, E., Patel, V. C., and Nash, J. F., "Mean Wind Profiles in Hurricanes," J. Eng. Mech. Div., ASCE, 102, No. EM2, Proc. Paper 12044, April 1976, pp. 265-273.

132. Sinclair, R. W., Anthes, R. A., and Panofsky, H. A., Variation of the Low Level Winds During the Passing of a Thunderstorm Gust Front, NASA Contractor Report No. Cr-2289, 1973.

133. Hinze, J. O., Turbulence, McGraw-Hill, New York, NY, 1959.

134. Shiotani, M., Structure of Gusts in High Winds, Interim Report, Parts 1-4, The Physical Sciences Laboratory, Nihon University, Furabashi, Chiba, Japan, 1967-1971.

135. Duchêne-Marullaz, P., "Effect of High Roughness on the Characteristics of Turbulence in the Case of Strong Winds," Proceedings, Fifth International Conference on Wind Engineering, Vol. 1, Colorado State Univ., Fort Collins, CO, Pergamon Press, 1980.

136. Teunissen, H. W., Characteristics of the Mean Wind and Turbulence in the Planetary Boundary Layer, Review No. 32, Institute for Aerospace Studies, University of Toronto, 1970.

137. Panofsky, H. A., Discussion to "Equivalent Static Wind Loads for Tall Buildng Design," by E. Simiu, Journal of the Structural Division, ASCE, April 1977, p. 925.

138. Simiu, E., Closure to "Equivalent Static Wind Loads for Tall Building Design," Journal of the Structural Division, ASCE, April 1977, p. 925.

139. Davenport, A. G., Discussion of "Equivalent Static Wind Loads for Tall Building Design" by E. Simiu, Journal of the Structural Division, ASCE, Feb., 1978, pp. 463-467.

140. Davenport, A. G., "The Dependence of Wind Load Upon Meteorological Parameters," Proceedings of the International Research Seminar on Wind Effects on Buildings and Structures, University of Toronto Press, Toronto, Canada, 1968, pp. 19-82.

141. Vickery, B. J., On the Reliability of Gust Loading Factors, Proceedings of the Technical Meeting concerning Wind Loads on Buildings and Structures, National Bureau of Standards, Building Science Series 30, Washington, DC, 1970, pp. 93-104.

142. Ropelewski, C. F., Tennekes, H., and Panofsky, H. A., "Horizontal Coherence of Wind Fluctuations," Boundary-Layer Meteorology, Vol. 5, pp. 11-20.

143. Brook, R. R., A Note on Vertical Coherence of Wind Measured in an Urban Boundary Layer," Boundary-Layer Meteorology, Vol. 9, 1975, pp. 247-256.

144. Panofsky, H. A., and Mizuno, T., "Horizontal Coherence and Pasquill's Beta," Boundary-Layer Meteorology, Vol. 9, 1975, pp. 247-256.

145. Berman, S., and Stearns, C. R., "Near-Earth Turbulence and Coherence Measurements at Aberdeen Proving Ground, Maryland," Boundary-Layer Meteorology, Vol. 11, 1977, pp. 485-506.

146. Kanda, J., Royles, R., "Further Consideration of the Height Dependence of the Root-Coherence in the Natural Wind," Building and Environment, Vol 13, 1978, pp. 175-184.

147. Stegen, G. R., and Thorpe, R. L., "Vertical Coherence in the Atmospheric Boundary Layer," Proceedings, Fifth International Conference on Wind Engineering, Vol. 1, Colorado State University, Fort Collins, CO, Pergamon Press, 1980.

148. Durst, C. S., "Wind Speeds Over Short Periods of Time," Meteorological Magazine, Vol. 89, 1960, pp. 181-186.

149. Court, A., "Wind Extremes as Design Factors," Journal of the Franklin Institute, Vol. 256, July 1953, pp. 39-55.

150. Simiu, E. and Filliben, J. J., "Probability Distributions of Extreme Wind Speeds," Journal of the Structural Division, ASCE, Vol. 102, September 1976, pp. 1861-1877.

151. Simiu, E., Bietry, J., and Filliben, J. J., "Sampling Errors in Estimation of Extreme Speeds," Journal of the Structural Division, ASCE, March 1978, pp. 491-501.

152. Surry, D. and Davenport, A. G., "Modeling the Wind Climate: An "Over View," Proceedings, Workshop on Wind Climate, Electrical Power Research Institute and National Science foundation Asheville, NC, 1978.

153. Thom, H. C. S., "New Distributions of Extrem Wind Speeds in the United States," Journal of the Structural Division, ASCE, Vol. 94, July 1968, pp. 1787-1801.

154. Riera, J. D., Viollaz, A. J., and Raimundin, J. D., "Some Recent Results of Probabilistic Models of Extreme Wind Speeds," Journal of Industrial Aerodynamics, Vol. 2, No. 3, November 1977, pp. 271-287.

155. Ito, M., Fujino, Y., "Design Wind Speed and Wind Load Factor Based on Probabilistic Rationale," Proceedings, Fifth Internaational conference on Wind Engineering, Fort Collins, CO, July 1979.

156. Simiu, E. and Filliben, J. J., "Weibull Distributions and Extreme Wind Speeds," Journal of the structural Division , ASCE, Jan., 1980, pp. 2365-2374.

157. Vickery, B. J., "On the Estimation of Extreme Speeds in Mixed Wind Climates," Proceedings, Workshop on Wind Climate, Electrical Power Research Institute and National Science Foundation, Asheville, NC, November 1979.

158. Simiu, E., Filliben, J. J., and Shaver, J. R., "Short - Term Records and Extreme Wind Speeds,". Journal of the Structural Division, ASCE, Vol. 108, No. ST11, November, 1982, pp. 2571-2577.

159. M. Grigoriu, "Estimation of Extreme Winds from Short Records," Journal of Structural Engineering, Vol. 110, No. 7, July, 1984, pp. 1467-1484.

160. Simiu, E. and Filliben, J. J., "Wind Direction Effects on Cladding and Structural Loads," Engineering Structures, Vol. 3, July 1981, pp. 181-186.

161. Davenport, A. G., "The Prediction of Risk Under Wind Loading," Proceedings, ICOSSAR Conference, Munich, 1977.

162. Wen, Y.K., "Wind Direction and Structural Reliability," Journal of Structural Engineerging, Vol. 109, April, 1983, pp. 1023-1041.

163. Wen, Y. K., "Wind Direction and Structural Reliability II," Journal of Structural Engineering, Vol. 110, June 1984, pp. 1253-1264.

164. Russell, L. R., "Probability Distributions for Hurricanes Effects" Journal of Waterways, Harbors, and Coastal Engineering Division, ASCE, Vol. 97, February 1971, pp. 139-154.

165. Meteorological Criteria for Standard Project Hurricane and Probable Maximum Hurricane Windfields, Gulf and East Coasts of the United States, NOAA Technical Report NWS 23, National Bureau of Standards National Oceanic and Atmospheric Administration, Washington, DC, 1979.

166. Holliday, C., On the Maximum Sustained Winds in Atlantic Hurricanes, Technical Memorandum No. WBTM-ST-45, U.S. Dept. of Commerce, Environmental Sciences Services Administration, Weather Bureau. Southern Region, Fort Worth, TX, 1969.

167. Tryggvason, B. V., Surry, D., and Davenport, A. G., "Predicting Wind-Induced Response in Hurricane Zones," Journal of the Structural Division, ASCE, Vol. 102, December 1976, pp. 2333-2350.

168. Gomes, L., and Vickery, B. J., On the Prediction of Tropical Cyclone Gust Speeds Along the Northern Australian Coast, Research Report R278, School of Civil Engineering, The University of Sydney, February 1976.

169. Tryggvasson, B. V., Computer Simulation of Tropical Cyclone Wind Effects for Australia, James Cook University of North Queensland, Dept. of Civil and System Engineering, Wind Engineering Report 2/79, Townsville, Queensland, Australia, April 1979.

170. Batts, M. E., Russell, L. R., and Simiu, E., "Hurricane Wind Speeds in the United States," Journal of the Structural Division, ASCE, October 1980.

171. Batts, M. E., Cordes, M. R., and Simiu, E., "Sampling Errors in Estimation of Extreme Hurricane Winds," Journal of the Structural Division, ASCE, October 1980.

172. Batts, M. E., et al., Hurricane Wind Speeds in the United States, BSS 124, National Bureau of Standards, Washington, DC, 1980.

173. Simiu, E., and Batts, M. E., "Wind-Induced Cladding Loads in Hurricane-Prone Regions," Journal of Structural Engineering, Vol. 109, January, 1983, pp. 262-266.

174. Hurricane-Induced Wind Loads, Computer Program, Accession Number No. PB 82132259, National Technical Information Service, Springfield, VA, 1982.

175. Simiu, E., Changery, M. J., and Filliben, J. J., Extreme Wind Speeds at 129 Stations in the Contiguous United States, National Bureau of Standards Building Science Series 118, March 1979 (Summary in Journal of the Structural Division, ASCE, March 1980).

176. Changery, M. E., et al., Directional Extreme Wind Speed Data for the Design of Buildings and Other Structures, Building Science Series BBS 160, National Bureau of Standards, Gaithersburg, MD, March, 1984.

177. Simiu, E., et al., "Multivariate Distributions of Directional Wind Speeds," Journal of Structural Engineering, Vol. 111, April, 1985, pp. 939-943.

178. Markee, E. H., Beckerley, J. G., and Sanders, K. E., Technical Basis for Interim Regional Tornado Criteria, WASH-1300 (UC-11), Nuclear Regulatory Commission, Washington, DC, 1974.

179. Abbey, R. R., "Risk Probabilities Associated with Tornado Wind Speeds," Proceedings, Symposium on Tornadoes, June 22-24, 1976, Texas Tech University, Lubbock, TX, pp. 177-236.

180. Wen, Yi-Kwei, and Chu, Shih-Lung, "Tornado Risk and Design Wind Speed," Journal of the Structural Division, ASCE, December 1973, pp. 2409-2421.

181. Twisdale, L. A., "Tornado Data Characterization and Wind Speed Risk," Journal of the Structural Division, ASCE, October 1978, pp. 1611-1630.

182. Schaeffer, J. T., Kelly, D. L., and Abbey, R. F., "Tornado Track Characteristics and Hazard Probabilities," Proceedings, Fifth International Conference on Wind Engineering, Vol. 1, 8-14, July 1979, Colorado State University, Fort Collins, CO.

183. Fujita, T. T., Workbook of Tornadoes and High Winds for Engineering Applications, Satellite and Mesometeorology Research Project, Department of Geophysical Sciences, The University of Chicago, SMRP Research Paper 165, September 1978.

184. Minor, J. E., McDonald, J. R., and Mehta, K. C., The Tornado: An Engineering Oriented-Prospective, NOAA Technical Memorandum ERL NSSL-82, National Severe Storms Laboratory, Norman, OK, December 1977.

185. Twisdale, L. A., and Dunn, W. L., Tornado Missile Simulation and Design Methodology, EPRI NP-2005, Electrical Power Research Institute, Palo Alto, CA., Aug., 1981.

186. Lew, M. and Hart, G. C., "Microzonation in Wind Engineering," Journal of the Structural Division, ASCE, Vol. 105, June 1979, pp. 975-989.

2. FUNDAMENTALS OF AERODYNAMICS, STRUCTURAL DYNAMICS, AEROELASTICITY AND WIND TUNNEL TESTING

Aerodynamics is concerned with the study of forces arising from the relative motion of a fluid with respect to a body. Aerodynamic forces induced on a structure by atmospheric flows generally vary with time. This variation is due to the state of turbulence of the oncoming wind flow, and to the local or "signature" flow fluctuations provoked in the wind by the structure itself. Because the aerodynamic forces are dependent on time, the methods of structural dynamics may have to be employed to determine the structural response. Indeed, if the fluctuating aerodynamic forces acting on a structure contain sufficiently strong harmonic components with frequencies equal or close to the structure's natural frequency (or frequencies), resonant amplification phenomena occur that must be taken into account in design for both strength and serviceability.

If the structure may be regarded for practical purposes as stiff, the relative motion of the fluid with respect to the structure is due only to the motion of the fluid itself. However, if the structure is flexible, the relative motion is due not only to the motion of the fluid, but to the flow-induced motion of the structure as well. A complex feedback mechanism thus occurs, whereby the aerodynamic forces that cause the structure to move are themselves modified by the motion of the structure. If this modification is such that the motions produced by the aerodynamic forces keep increasing with time, an aeroelastic instability is said to occur. Aeroelasticity is concerned with the study of aerodynamic forces and structural motions significantly influenced by the feedback mechanism just described.

With rare exceptions - which are primarily of scientific, rather than immediate practical interest - the wind flow around a bluff body and the pressures exerted by the flow at each point of the body cannot be determined from first principles. Information on aerodynamic pressures (or forces or moments) must therefore be based on measurements. For practical reasons most aerodynamic data are- and have been in the past - obtained by wind tunnel measurements. Full-scale measurements are only conducted to provide information that cannot be obtained reliably in the laboratory, and benchmark data against which wind tunnel performance can be assessed.

The objective of this section is to provide basic notions of aerodynamics, structural dynamics, aeroelasticity, and wind tunnel testing. It is the Committee's view that an understanding of these notions is a valuable asset to the designer of wind-sensitive structures. More detailed information on the topics covered here will be provided in subsequent sections dealing with wind effects on specific types of structure.

2.1 FUNDAMENTALS OF AERODYNAMICS

2.1.1 Basic Principles and Relations

2.1.1.1 Principle of Conservation of Mass.

This principle states that, during any given time interval, the mass of fluid entering a given volume is equal to the mass of fluid leaving that volume. Consider fig. 2.1.1, in which the heavy lines represent streamlines. By virtue of the principle of conservation of mass, flow speeds will be higher through the constricted area A_2 than through the area A_1. Similarly, in fig. 2.1.2, wind speeds along the streamlines 1′ -2′ and 1" -2" are higher at points 2′ and 2" than at points 1′ and 1", respectively.

2.1.1.2 Bernoulli's Equation. By virtue of Newton's second law, an elemental air particle in a wind flow is subjected to a force of inertia proportional to its acceleration. This force is balanced, in general, by forces associated with (a) normal pressures, and (b) shear stresses due to the air viscosity. In steady flow and in regions where the shear stresses are negligible, the following relation can be obtained:

$$\frac{1}{2}\rho u^2 + p = \text{const} \qquad (2.1.1)$$

where ρ = air density, u = velocity, p = pressure. Eq. 2.1.1 is a special form of Bernoulli's theorem.

2.1.1.3 Dynamic Pressures. The quantity

$$q = \frac{1}{2}\rho u^2 \qquad (2.1.2)$$

has the dimensions of pressure and is referred to as the dynamic pressure at the point under consideration. For standard air [0.07651 lb per cu ft (12.00717 N/m^3), corresponding to 15^oC and 760 mm of mercury] and velocity u expressed in miles per hour, the dynamic pressure in pounds per square foot is given by

$$p = 0.002558\, u^2 \qquad (2.1.3)$$

If u is expressed in meters per second, the dynamic pressure in Pascals is given by

$$q = 0.612\, u^2 \qquad (2.1.4)$$

2.1.1.4 The Reynolds Number. The characteristics of a fluid flow depend to a large extent upon the relationship between inertial and viscous forces within the flow. A nondimensional measure of this relationship is given by the Reynolds number

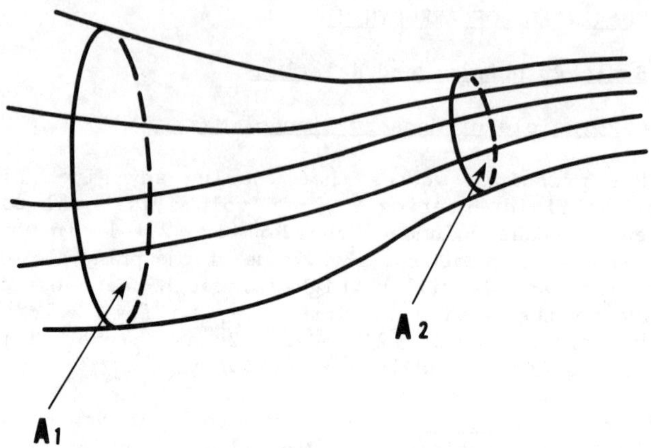

Figure 2.1.1 Streamlines around two areas, A_1 and A_2.

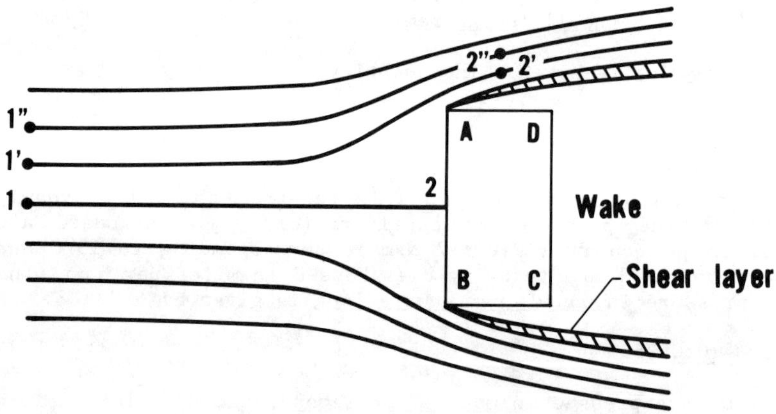

Figure 2.1.2 Streamlines around rectangular shape.

$$Re = \frac{u\,D}{\nu} \qquad (2.1.5)$$

where u = characteristic speed, D = characteristic body dimension, and ν = kinematic viscosity. For air at 15°C, $\nu \simeq 1.5 \times 10^{-5}\ m^2/s$, so that

$$Re \simeq 67{,}000\ u\,D \qquad (2.1.6)$$

where u and D are expressed in meters per second and meters, respectively.

2.1.2 Flow Around a Bluff Body

2.1.2.1 Pressure and Suction Zones. A typical flow around a bluff body with sharp edges immersed in a uniform flow is shown in fig. 2.1.2, in which the heavy lines represent streamlines. It is seen that in this case flow separation occurs at the corners A and B of the windward face. Shear layers originating from the separation points surround a region referred to as the wake.

Along the streamline 1-2 (fig. 2.1.2), where 1 is a point far upwind of the body, an elemental air mass will be decelerated as it approaches point 2. When the elemental mass reaches point 2 (referred to as the stagnation point), its speed vanishes. The pressure at point 2 can then be calculated from Bernoulli's equation (Eq. 2.1.1):

$$p_2 = \frac{1}{2}\rho u_1^2 + p_1$$

At point 1, the flow speed u_1 is equal to the undisturbed flow speed upwind of the structure, denoted by u_∞, and the pressure p_1 is equal to the atmospheric pressure, p_a. Since $u_2 = 0$, it follows that the pressure at point 2 is higher than the atmospheric pressure p_a, an intuitively obvious result. For the body of fig. 2.1.2 pressures are higher than p_a throughout the windward face AB.

Let now Eq. 2.1.1 be written for streamline 1'-2' (point 1' being again assumed to be far upstream of the body):

$$\frac{1}{2}\rho u_{2'}^2 + p_{2'} = \frac{1}{2}\rho u_\infty^2 + p_a$$

It follows from the preceding relation and the inequality $u_{2'} > u_\infty$ (see section. 2.1.1.1) that $p_{2'} - p_a < 0$, i.e., the pressure at and near point 2' is lower than the atmospheric pressure. Zones on the surface of the body where pressures are lower than p_a (i.e., negative pressure, or suction, zones) include in the case of fig. 2.1.2 the sides AD, BC, and the leeward face CD.

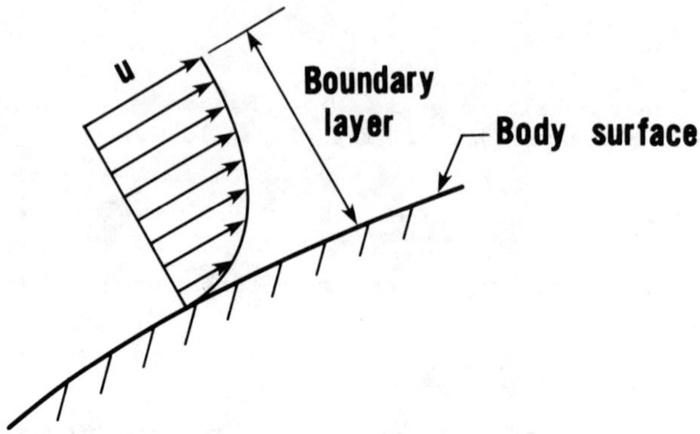

Figure 2.1.3 Boundary layer near a body surface

2.1.2.2 Flow Separation Mechanisms. It is an experimental fact that, in an air flow, the air in contact with a body will adhere to its surface. This causes a retardation of the air motion within a thin zone, known as boundary layer, that forms near the body surface. Within the boundary layer the mean speed varies from zero at the body surface to its full value, u, corresponding to the external (as opposed to boundary layer) flow (fig. 2.1.3).

Certain regions adjacent to the body surface are characterized by negative pressure gradients, i.e., pressures increasing as one moves downward. In these regions the boundary layer flow at a given distance from the surface is decelerated. If the pressure gradients are sufficiently strong, flow separation occurs. In flows around bodies with sharp corners, separation will occur at corners, where negative pressure gradients are strong (figs. 2.1.2, 2.1.4).

For bodies with a curved shape in plan, the location of the separation points depends upon factors that include body shape and nature of the flow within the boundary layer. Separation will occur earlier (i.e., near the stagnation point) if the boundary layer is laminar, and farther downwind from the stagnation point if the boundary layer is turbulent (figs. 2.1.5a and 2.1.5b). This can be explained as follows. In the regions of the boundary layer farthest removed from the body, the retardation of the flow is weak and fluid particles have, therefore, relatively high kinetic energy. As a result of turbulent mixing within the boundary layer, such particles are transported toward the body surface, where their ability to counteract the effects of negative pressure gradients is higher than would be the case if the boundary layer flow were laminar; separation will thus occur farther downwind.

2.1.2.3 Wake Flows. Near separation zones strong shear stresses impart to the fluid particles rotational motions that generate

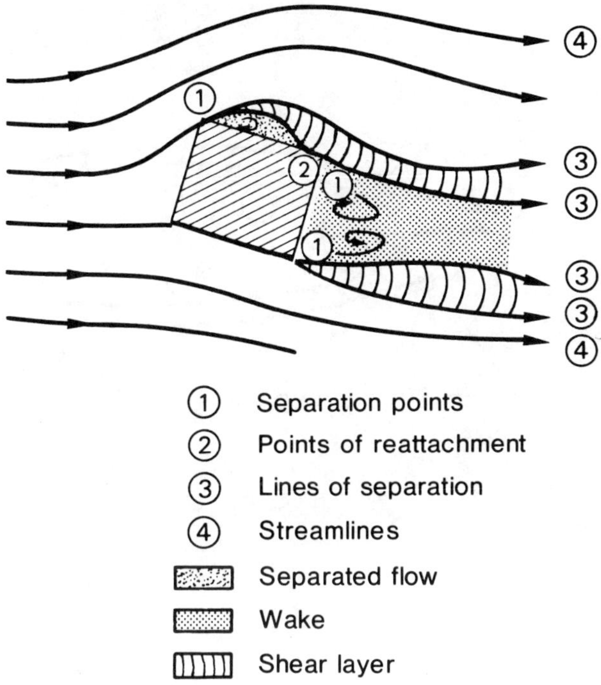

a

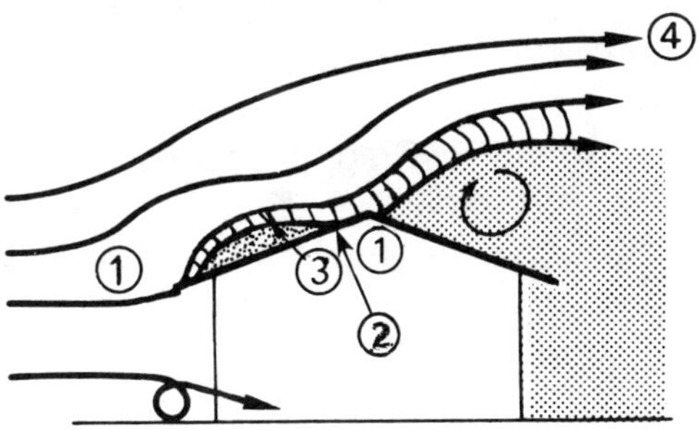

b

Figure 2.1.4 Zones in flow around bodies with sharp corners. From Aérodynamique, Centre Scientifique et Technique du Bâtiment, Nantes, France, 1980.

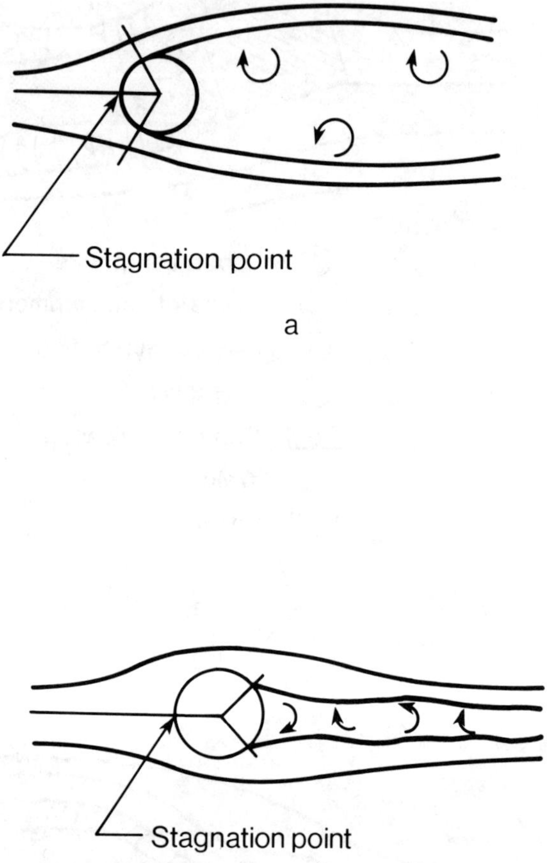

Figure 2.1.5 Separation points on circular cylinder when boundary layer over cylinder surface is (a) laminar and (b) turbulent.

discrete vortices in the separation layers. Through mechanisms that are not well understood, vortices are shed in the wake of the body, which are more or less well organized depending upon the Reynolds number of the flow. Well organized sequences of alternating vortices are referred to as von Kármán (or Bénard) vortex streets. Flow visualizations of vortex streets are available in the literature for both laboratory and geophysical flows -- the latter in the form of satellite pictures of cloud formations in the wake of islands with quasi-cylindrical shapes (Ref. 202, pp. 137, 140 and 141).

2.1.2.4 <u>The Strouhal Number</u>. The frequency, n_s, at which vortices are then shed in the wake of cylindrical body was found experimentally to satisfy the relation

$$S = \frac{n_s D}{u} \tag{2.1.7}$$

where S = Strouhal number and D = typical cross-flow dimension. The frequency, n_s, is referred to as the Strouhal frequency. The Strouhal number generally depends upon the cross-sectional shape of the cylindrical body, the Reynolds number Re, the turbulence characteristics of the oncoming flow, and the details of the exterior surface (roughness, texture, presence or absence of ribs).

For a long circular cylinder with a smooth surface immersed in uniform flow the relation between Strouhal number and Reynolds number is shown for $Re > 10^5$ or so in fig. 2.1.6 [201]. Note that the shedding of vortices is labeled in fig. 2.1.6 as periodic in the lower Reynolds number range, regular in the higher Reynolds number range, and random at $3 \times 10^5 < Re < 3 \times 10^6$, where as a result of transition from laminar to turbulent boundary layer flow the vortex shedding occurs over a relatively broad range of frequencies centered about n_s.

In the case of a prismatic shape with sharp corners, it is widely believed - though not yet proven experimentally - that the vortex shedding phenomenon is independent of Reynolds number.

2.1.3 <u>Pressure, Drag, Lift, and Moment Effects</u>

2.1.3.1 <u>Aerodynamic Coefficients</u>. Each point on the surface of a body experiences pressures due to the action of the air flow. It is usual to refer all pressures to the mean dynamic pressure at some reference point where the flow is undisturbed by the presence of the body. Nondimensional <u>pressure coefficients</u> are thus defined as follows:

$$C_p = \frac{p - p_o}{q} \tag{2.1.8}$$

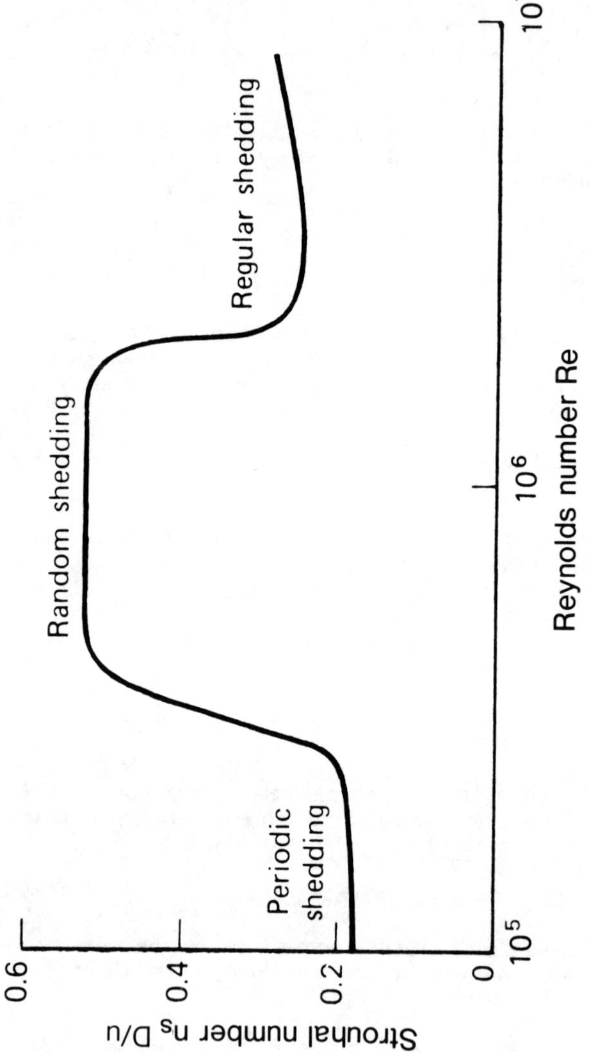

Figure 2.1.6 Dependence of Strouhal number upon Reynolds number, flow around circular cylinder.

where $q = 1/2\,\rho u^2$ is the dynamic pressure at the reference point (u = mean speed, p = local pressure, and p_o = pressure in the undisturbed flow far upwind of the body).

The resultant aerodynamic force acting on the body is usually resolved into two components, one parallel and the other normal to the direction of the mean speed in the undisturbed wind flow. These are referred to as the drag (or along-wind) force, and the lift (or across-wind) force, respectively. The aerodynamic moment with respect to some arbitrary point is equal to the product of the aerodynamic force by its moment arm with respect to that point.

Nondimensional drag, lift, and aerodynamic moment coefficients are defined as follows:

$$C_D = \frac{D}{qA} \qquad (2.1.9)$$

$$C_L = \frac{L}{qA} \qquad (2.1.10)$$

$$C_M = \frac{M}{qAB} \qquad (2.1.11)$$

where D = drag force, L = lift force, M = aerodynamic moment, A = exposed area, B = some conveniently chosen dimension.

As previously indicated, pressures, drag and lift forces, and aerodynamic moments, and therefore the aerodynamic coefficients C_p, C_D, C_L, C_M, must be determined experimentally. Note that if the aerodynamic loads fluctuate as a consequence of oncoming or body-induced turbulence, the above quantities are functions of time. In such cases mean, peak, r.m.s., probability distributions, and spectral densities of the aerodynamic coefficients may be defined.

The drag of a rectangular cylinder depends upon the ratio between height and width of its cross-section. If the ratio B/H (fig. 2.1.7) is larger than 0.5 or so flow reattachment occurs and the drag coefficient drops [202, 205].

In the case of a circular cylinder with a smooth surface, the pressure distribution depends upon the characteristics of the flow around the cylinder and is therefore a function of Reynolds number, as shown in section 2.1.2.2.

This is also true of the drag coefficient (fig. 2.1.8). Note that the dependence of both pressure distribution and drag force upon Reynolds number will change if the surface of the cylinder is rough or ribbed: indeed, the roughness, or the ribs, will cause the boundary layer to be turbulent -- and the structure of the wake

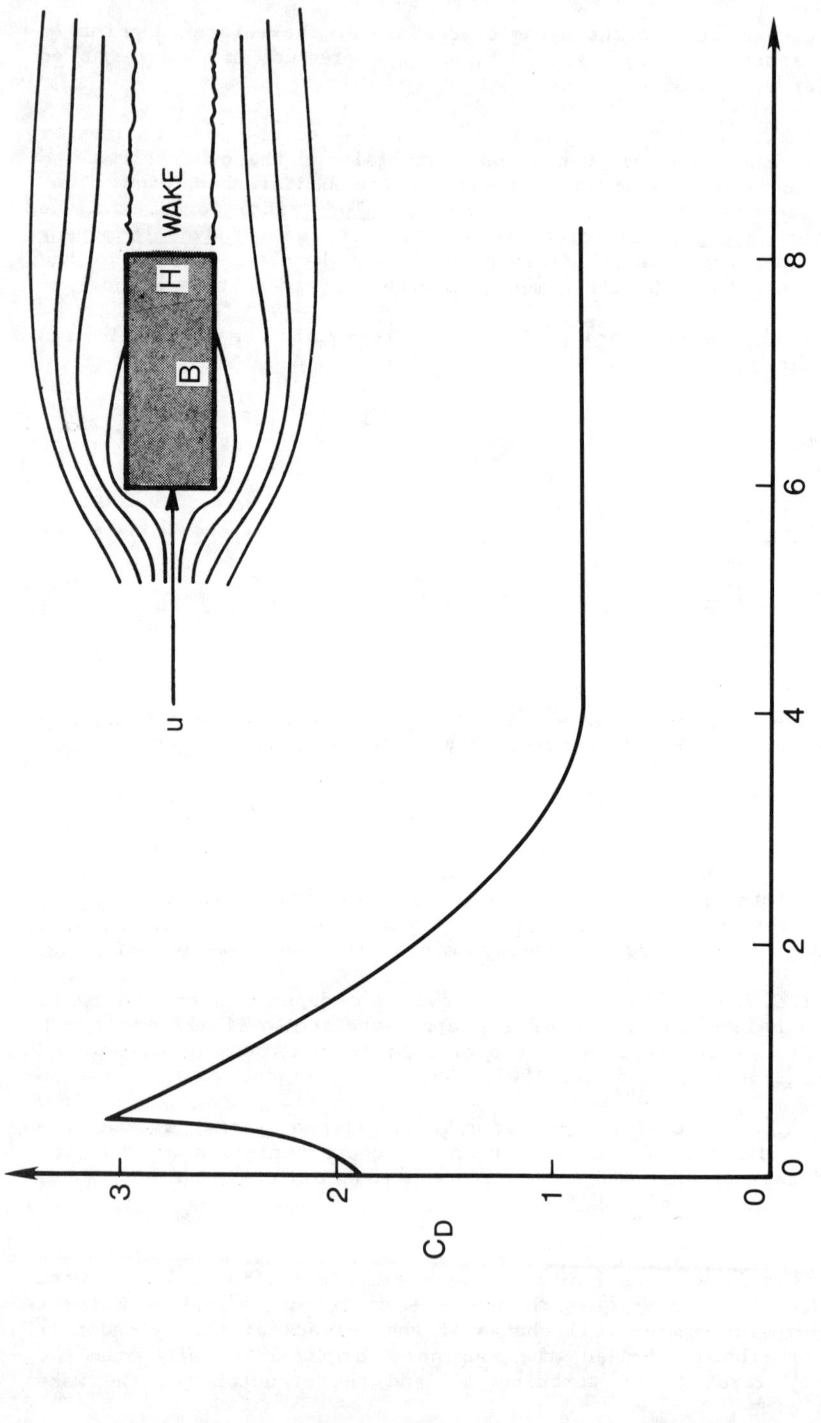

Figure 2.1.7 Dependence of drag on rectangular cylinder upon ratio B/H.

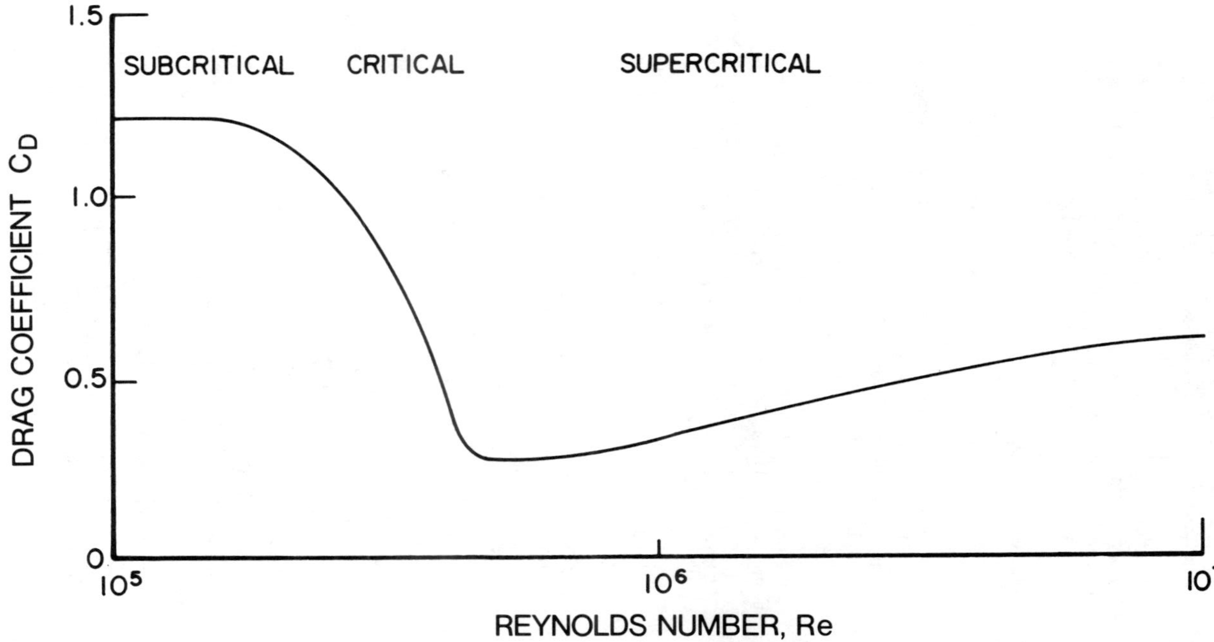

Figure 2.1.8 Dependence of drag coefficient upon Reynolds number, flow around circular cylinder.

flow to change -- at lower Reynolds numbers than would be the case if the surface of the cylinder were smooth. The observation that the drag can be reduced in certain Reynolds number ranges through the provision of surface ribs is applied, for example, to the design of golf balls and hyperbolic cooling towers (see Chapter 8). The provision of surface roughness, tripping wires, or ribs is also used extensively in wind tunnel testing [202, 206].

2.1.3.2 <u>Three-Dimensional and Turbulence Effects</u>. The drag coefficient for a very long square cylinder in uniform, smooth flow is C_D 2.0 [202, 204]. However, if the aspect ratio of the cylinder is of the order of 10 or less, the flow may no longer be assumed to be for practical purposes two-dimensional, and experiments show that end flow effects cause a significant reduction of the overall drag. For square building models standing on a wind tunnel floor in uniform, smooth flow, C_D 1.3 [208].

The drag coefficient is also influenced by turbulence in the oncoming flow. For example, if bodies with a curved shape in plan are immersed in turbulent flows, drag force reductions similar to those illustrated in fig. 2.1.8 occur at lower Reynolds number than would be the case if the oncoming flow were smooth.

In the case of a rectangular cylinder immersed in turbulent flow the mean drag forces depend upon the ratio of the width B to the depth H [202, 205].

Fig. 2.1.9 published in 1965 by Jensen and Franck [209] shows measurements of mean pressure coefficients for a house in (a) atmospheric flow, i.e., turbulent flow with height-dependent mean wind speeds as described in sections 1.2.1 and 1.2.2 (interrupted lines), and (b) in uniform, smooth flow (solid lines). Note that differences between wind loads in the two flow regimes are significant.

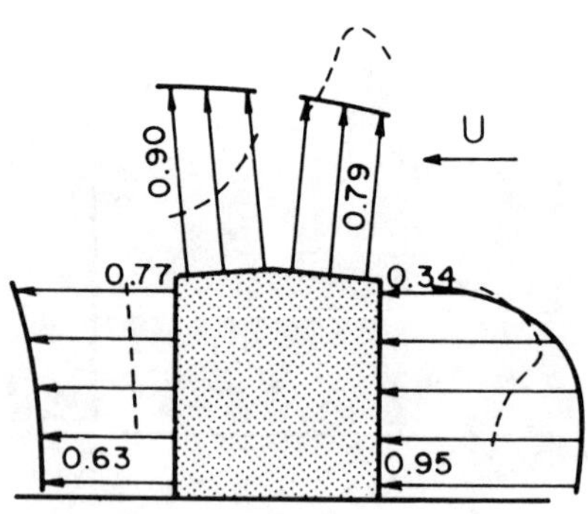

Figure 2.1.9 Mean Pressure Coefficients on Building in Uniform and Boundary Layer Flow

In addition to affecting mean loads, as shown previously, turbulence in the oncoming flow has the following effects. First, it induces time-varying loads, commonly referred to as buffeting loads, which are directly caused by the velocity fluctuations in the oncoming flow. Second, it may modify -- in certain instances dramatically -- time-dependent loads induced by "signature" turbulence (i.e., fluctuations provoked in the flow by the body itself). Both effects will be dealt with briefly in section 2.1.3.3.

2.1.3.3 Time-Varying Loads. A time-varying load, $F(t)$, may be described as a sum of a mean load, $\overline{F}(t)$, and a fluctuating load with zero mean, $F'(t)$:

$$F(t) = \overline{F}(t) + F'(t) \tag{2.1.12}$$

$F'(t)$ may be characterized by its root mean square (r.m.s.),

$\overline{F'^2(t)}^{1/2}$, by its peak value, Fpk, and, in certain applications, by its probability distribution, or by its spectrum, $S_{F'}(n)$. In the case of two loads, $F_1(t)$ and $F_2(t)$, acting at a distance, r, from each other, the extent to which their fluctuating parts act in phase or at cross-purposes is usefully characterized by the cross-spectrum $S_{F_1'F_2'}(r,n)$. (For the physical interpretation of the notions of spectrum and cross-spectrum, see sections 1.2.2.3 and 1.2.2.4).

It is convenient in applications to define peak and r.m.s. coefficients for pressures, drag forces, and so forth. For example, the peak pressure coefficient is

$$C_p^{pk} = \frac{p_{pk} - p_o}{q} \tag{2.1.14}$$

where the definitions of p_o and q are given in section 2.1.3.1.

If the area under consideration is small (i.e., a few square meters or less), the peak force (or area load), F_A^{pk} acting on an area, A, may be assumed to be approximately equal to the product of that area by the largest peak pressure acting at some point within A, i.e.,

$$F_A^{pk} = p_{pk}A \tag{2.1.15}$$

However, for larger areas, owing to the spatial non-uniformity of pressures,

$$F_A^{pk} < p_{pk}A \tag{2.1.16}$$

This inequality becomes stronger as the area A becomes larger.

The magnitude of F_A^{pk}, or of the corresponding force coefficient, as a function of A can generally be determined only by measurement. However, if the mean wind speed is normal to the windward face of a rectangular structure, the buffeting forces over that face can be obtained from a knowledge of the characteristics of the longitudinal velocity fluctuations. The latter are obtained from the relation

$$P(Q,t) \approx \frac{1}{2}\rho[u(z) + u'(z,t)]^2 C_p(Q) \tag{2.1.17}$$

where P(Q,t) = total pressure (i.e., sum of mean and fluctuating pressure). Since in the atmosphere the quantity $[u'(z,t)/u(z)]^2$ is much smaller than unity, there follows:

$$p'(Q,t) \approx \rho u(z)\, u'(z,t)\, C_p(Q) \tag{2.1.18}$$

where p'(Q,t) = fluctuating pressure at point Q on the windward face, ρ = air density, u(z) = mean wind speed, u'(z,t) longitudinal velocity fluctuation, z = elevation of point q, t = time, $C_p(Q)$ =mean pressure coefficient at point Q.

While drag forces can be affected significantly by the turbulence intensity, it appears that the effect of turbulence scale on drag is relatively weak [205, 206, 210]. However, turbulence scales may affect area loads significantly (see section 1.2.2.1) [206]. These considerations are important for wind tunnel testing purposes.

Fluctuating lift forces are due primarily to the asymmetry of the flow field as vortices are shed in the structure's wake. In the case of a cylindrical body immersed in uniform, smooth flow, the wake flow is approximately two-dimensional, so that the lift forces act coherently along the length of the body. If the oncoming flow is turbulent, some of this coherence is lost, and the overall lift force acting on the body decreases [211].

The case will now be considered of a tower immersed in an atmospheric flow (fig. 2.1.10). If vortices were shed independently at each elevation, the vortex shedding frequency at elevation z would be

$$n_s(z) = \frac{u(z)\cdot S}{D} \tag{2.1.19}$$

(see Eq. 2.1.7). The tower would then be subjected to elemental lift forces with frequencies varying from zero to u(h)·S/D. On the other hand, if the flow were uniform, the frequency of the elemental lift forces throughout the tower length would be constant and equal

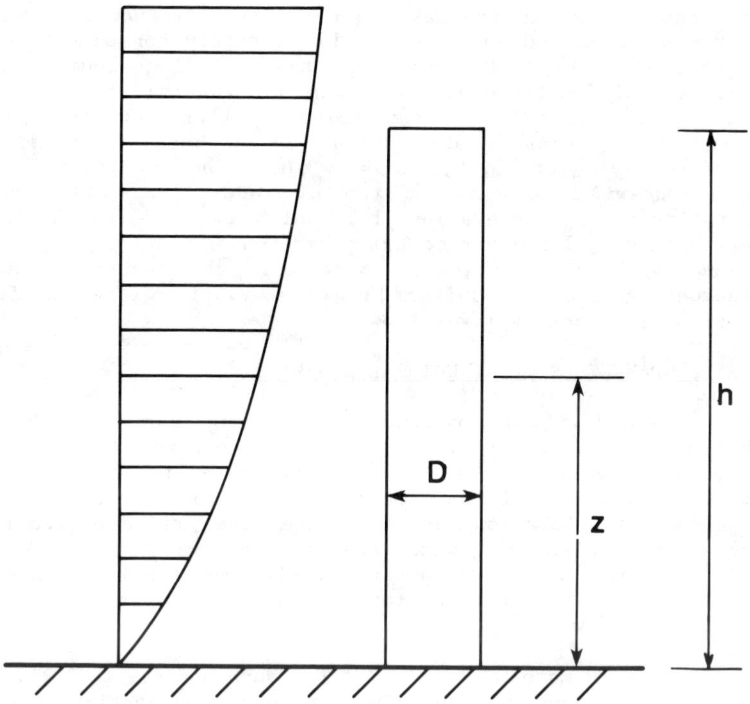

Figure 2.1.10 Tower immersed in atmospheric boundary layer flow.

or close to uS/D, where u = mean speed of the uniform flow. Thus, while in uniform flow the spectrum of the lift force would be relatively narrow, in an atmospheric flow the lift spectrum would according to Eq. 2.1.19 be broad.

In reality, Eq. 2.1.19 does not hold, i.e., the frequencies of vortices shed in the wake do not vary continuously. Rather, a vortex shedding pattern with cells of nearly constant frequencies is observed [212]. Nevertheless, the preceding argument explains qualitatively differences between fluctuating lift forces in atmospheric and uniform flows. Figure 2.1.11 shows the normalized flow-induced across-wind response of a model in uniform and boundary layer (or <u>shear</u>) flows, σ_y/H, versus $u_g n_y/D$ (σ_y = r.m.s. of across-wind response at top of model, u_g = flow speed at elevation z_g, z_g = elevation of top of boundary layer, H = building height ($H < z_g$), D = horizontal dimension of building, n_y = natural frequency in y direction). Note that the response induced by fluctuating lift in uniform flow is vastly different from the response in boundary layer flow.

2.2 FUNDAMENTALS OF STRUCTURAL DYNAMICS

The old military regulation requiring that troops break step when marching across a bridge is a well-known example of concern with structural dynamics effects. Its rationale lies in the fact that if the frequency of the steps is equal or close to a natural frequency of vibration of the bridge, the bridge deflection will increase cumulatively with each successive step. A similar resonance effect may occur in certain wind-sensitive structures. In this case, the exciting force is induced by speed fluctuations in the air flow.

A simple mathematical model for such phenomena is the spring, mass, dashpot system. The time-dependent load acting on the mass induces motion which is opposed by the restoring force due to a spring, and a damping force due to energy dissipation. This model will be dealt with briefly in sections 2.2.1 and 2.2.2. For additional details, see Ref. 202.

2.2.1 Single Degree of Freedom Linear System with Viscous Damping

For this case it is assumed that the mass is concentrated at a point, the spring is linearly elastic, and the damping inherent in the system is viscous, i.e., proportional to the speed of motion of the mass. The equation of motion of the system is

$$\ddot{x} + 2\,(2\pi n_1)\,\zeta_1\,\dot{x} + (2\pi n_1)^2\,x = \frac{F(t)}{m} \tag{2.2.1}$$

where n_1 = natural frequency, ζ_1 = damping ratio, m = mass, and F(t) = force acting on the system.

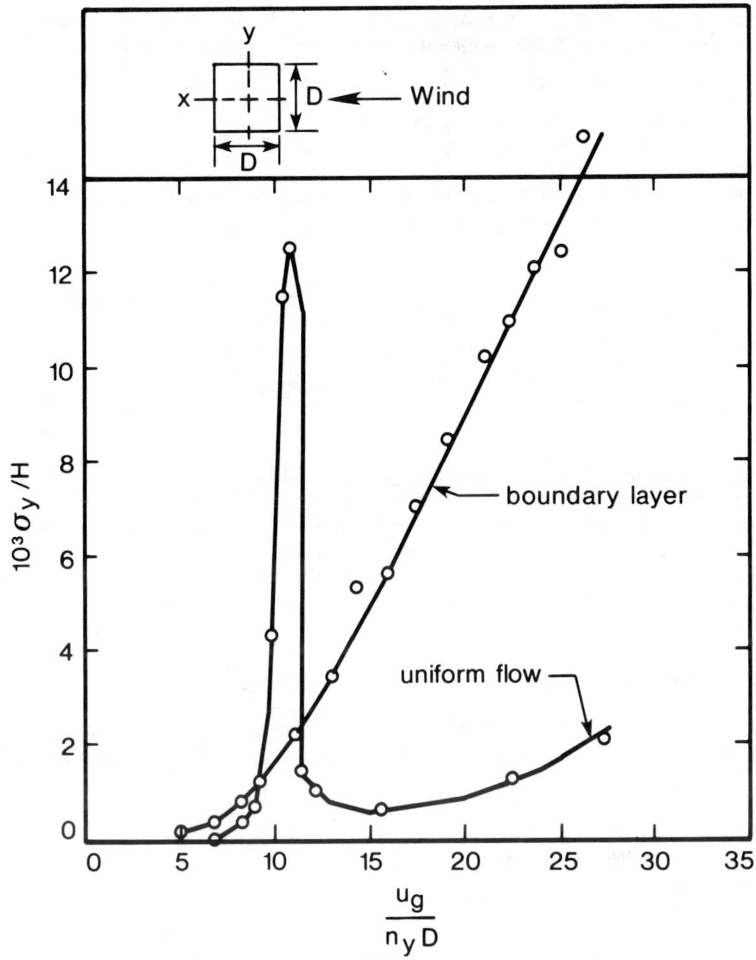

Figure 2.1.11 Across-wind response of a square tower in uniform and boundary-layer flow [213].

2.2.1.1 Response to a Harmonic Load. If F(t) is a harmonic force of frequency n, the displacement x(t) can be written as

$$x(t) = F_o H_1(n) \cos(2\pi nt - \phi) \tag{2.2.2}$$

where t = time, ϕ = phase angle, and $H_1(n)$ = mechanical admittance function, given by the expression

$$H_1(n) = \frac{1}{4\pi^2 n_1^2 m \{[1 - (n/n_1)^2]^2 + 4 \zeta_1^2 (n/n_1)^2\}^{1/2}} \tag{2.2.3}$$

Fig. 2.2.1 shows the variation of $(4\pi^2 n^2 m)^2 H^2(n)$ with n/n_1 for $\zeta_1 =$ 0.01.

2.2.1.2 Response to a Random Load. Loads occurring in wind engineering problems are rarely harmonic. Indeed -- whether they are generated by wind gustiness or by vorticity shed in the wake of the structure - wind loads generally vary more or less randomly with time. It is possible and convenient to represent such random loads as superpositions of elemental harmonic components, each with a different frequency. The force spectrum provides a measure of the frequency content of the force, i.e., of the magnitude of the harmonic components corresponding to each frequency.

If the system is acted upon by a random force with spectrum $S_F(n)$, the spectrum of the response, $S_x(n)$, can be calculated from the expression

$$S_x(n) = H_1^2(n) S_F(n) \tag{2.2.4}$$

Consider a system with $\zeta_1 = 0.01$, subjected to an aerodynamic force acting at the stagnation point of a building and having the spectrum, $S_F(n)$ [Fig. 2.2.2]. The spectrum of the displacement, $S_x(n)$, calculated by using Eq. 2.2.4, is also shown in Fig. 2.2.2. The shape of the spectrum $S_x(n)$ shows that the displacement is a random function dominated by harmonic components with frequencies equal or close to n_1. Typical records of the force F(t) and the displacement x(t) corresponding to the spectra $S_F(n)$ and $S_x(n)$ of fig. 2.2.2 (n_1 = 1sec), are shown in Figs. 2.2.3 and 2.2.4, respectively. An inspection of these records shows that the frequency content of F(t) is indeed richer than that of x(t), the system with response x being in effect a mechanical filter of the force input.

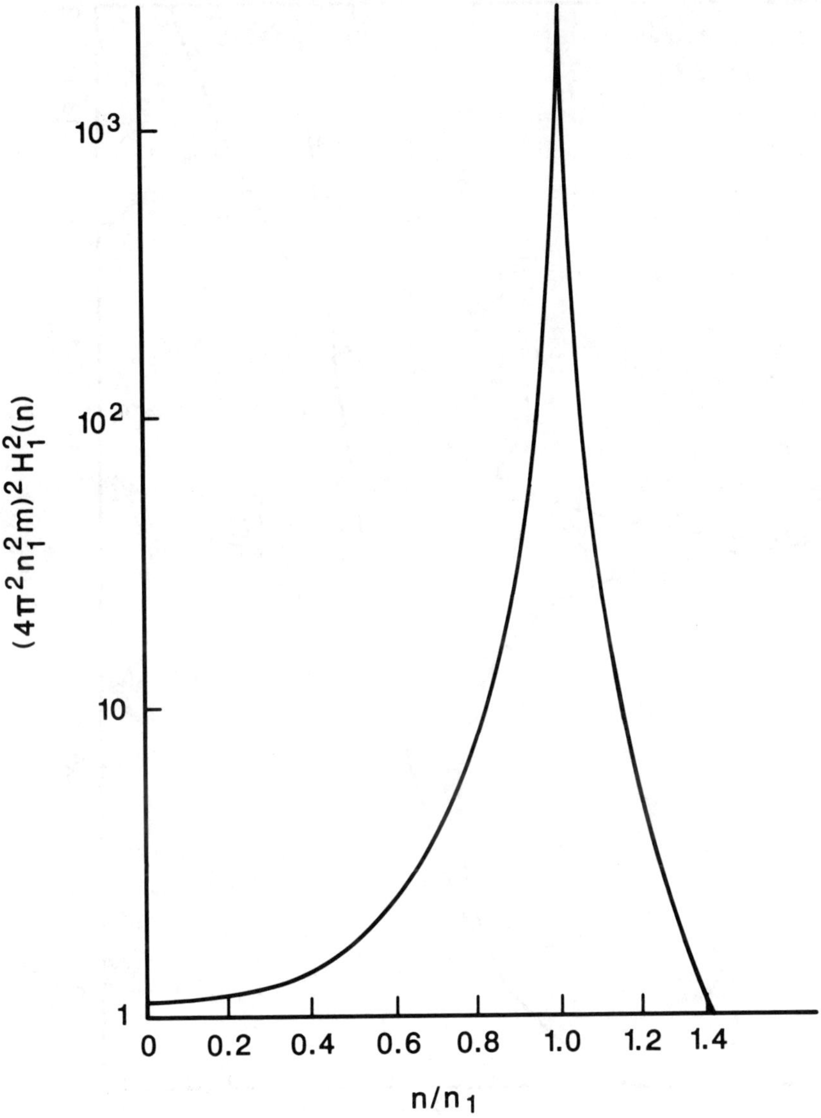

Figure 2.2.1 Function $H_1^2(n)$.

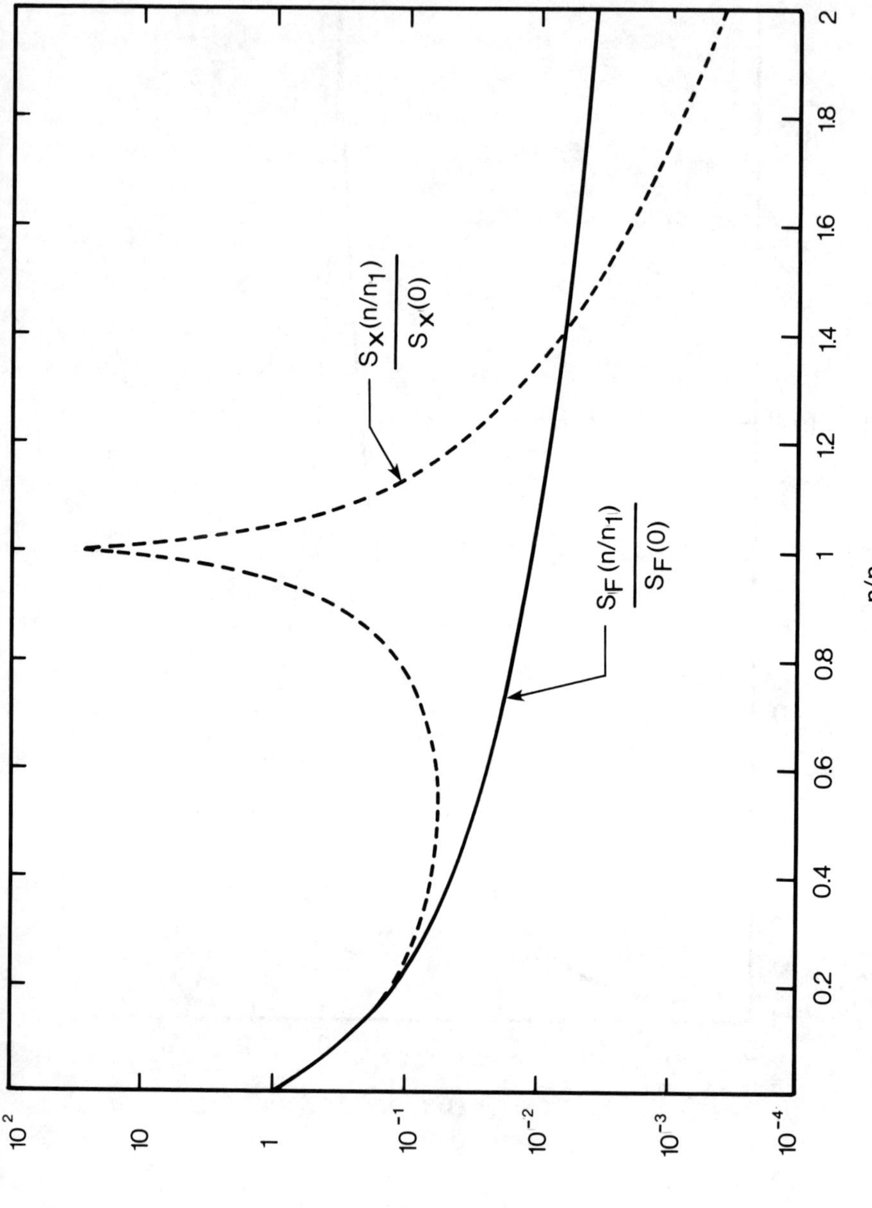

Figure 2.2.2 Input (force) and output (response) spectrum.

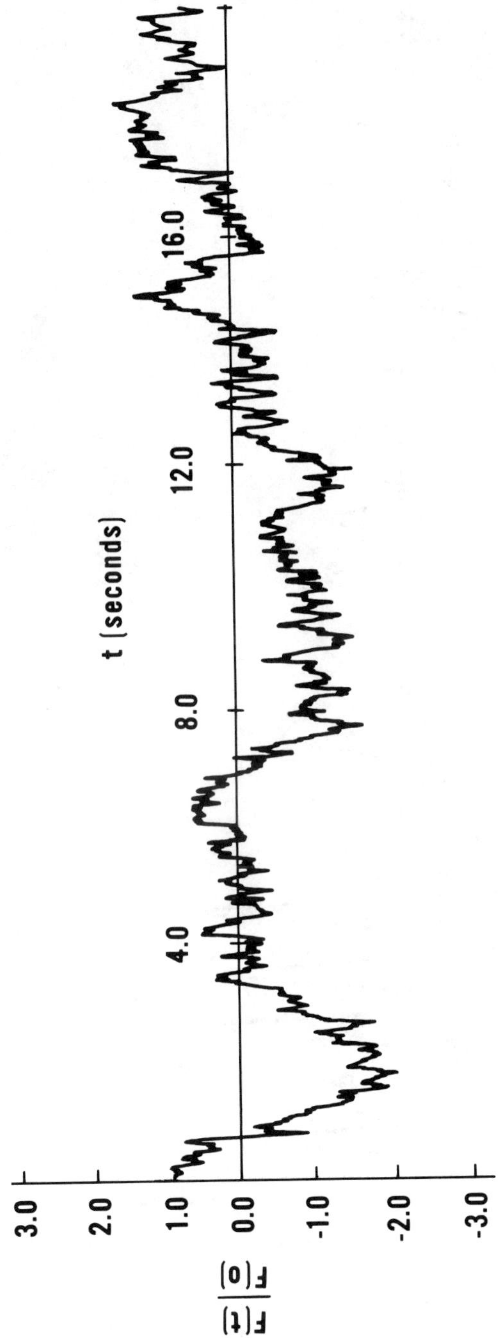

Figure 2.2.3 Record of aerodynamic force F(t).

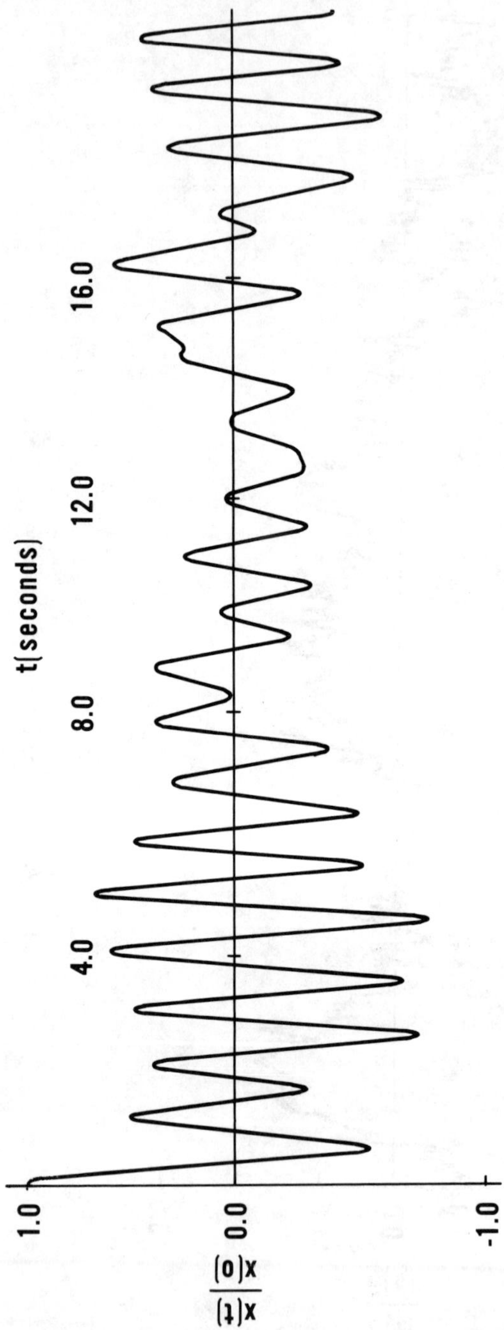

Figure 2.2.4 Record of Response x(t) to Force F(t).

2.2.2 Multi-Degree of Freedom Linear System with Viscous Damping

2.2.2.1 General Expression for Response of a Continuous Structure.

The deflections of continuous, linearly elastic systems with low damping can be written approximately in the form

$$x(z, t) = \sum_i x_i(z)\ \xi_i(t) \tag{2.2.5}$$

where z = running coordinate (e.g., height above ground in the case of a vertical structure), $x_i(z)$ = i-th normal mode of vibration, $\xi_i(t)$ = generalized coordinate in i-th mode (or i-th modal participation factor).

The normal modes, $x_i(z)$, are functions of the structure's mass and stiffness distribution, and are independent of the loads. Associated with each normal mode $x_i(z)$, is a natural frequency of vibration in the i-th mode, n_i, and a damping ratio, ζ_i. For details on procedures for determining normal modes and natural frequencies, see, e.g., Ref. 214. Modal damping ratios, ζ_i, can in principle be determined experimentally [215].

Each generalized coordinate $\xi_i(t)$, can be shown to represent the displacement of a one-degree-of-freedom system with frequency n_i, damping ratio ζ_i, and mass M_i, subjected to a force $Q_i(t)$. M_i and $Q_i(t)$ are referred to as the i-th mode generalized mass and generalized force, respectively, and have the expressions

$$M_i = \int_0^H x_i^2(z)\ m(z)dz \qquad (i=1,2,....) \tag{2.2.6}$$

$$Q_i = \int_0^H p(z,t)\ x_i(z)\ dz \qquad (i=1,2,....) \tag{2.2.7}$$

where H = length of structure, m(z) = mass of structure per unit of length, and p(z,t) = force per unit length acting on structure.

The preceding expressions apply to line-like structures. Similar expressions can be derived for three-dimensional structures experiencing longitudinal, transverse, and rotational displacements.

2.2.2.2 Response to a Distributed Random Load.

If the surface A of a structure is acted upon by a distributed random load, the spectrum of the displacement can be written as:

$$S_x(z,n) \simeq \sum H_i^2(n)\ x_i^2(z) \int_A \int_A x_i(z_1)x_i(z_2)S^{C'}_{p_1'p_2'}(n)dA_1dA_2 \tag{2.2.8}$$

where $H_i(n)$ is given by Eq. 2.2.3 in which the index 1 is everywhere replaced by the index i and m is replaced by M_i, and $S^C_{p'_1 p'_2}(n)$ is the real part of the cross-spectrum of the fluctuating pressures $p'_1(t)$ and $p'_2(t)$ acting at points P_1 and P_2 of the area A, respectively. Recall that for any given n, as the distance between P_1 and P_2 increases, the mutual coherence of the components with frequency n of the pressure loads p_1 and p_2 decreases. Note that in the case in which the mean wind speed is normal to the windward face of a rectangular structure, the fluctuating pressures are proportional to the fluctuating speeds (Eq. 2.1.18). Hence the cross-spectra of fluctuating pressures are in this case nearly proportional to the cross-spectra of longitudinal velocity fluctuations (Eq. 1.2.21). Eq. 2.2.8 is valid if the damping ratios ζ_i are small and the frequencies n_i (i=1,2,...) are well separated.

By virtue of the definition of the spectra, the mean square value of the fluctuating displacement, x'(z), induced by the fluctuating load is

$$\overline{x'^2(z)} = \int_0^\infty S_x(z,n)dn \tag{2.2.9}$$

The expected value of the peak displacement is

$$x'_{pk}(z) = K_x(z) \, \overline{x'^2(z)}^{1/2} \tag{2.2.10}$$

where for preliminary calculations it may be assumed that the peak factor $K_x(z) \simeq 3.5$ in most wind engineering applications.

2.2.3 Nonlinear Systems Under Random Loads

The response of a system with geometric or material nonlinearities subjected to random loads can be calculated numerically. Given the spectral and cross-spectra characterizing the loads, it is possible to simulate these loads as functions of time using techniques based on modeling by trigonometric function [216], on the more efficient Fast Fourier Transform approach [217], or on the so-called ARIMA methods [218].

The simulated loads are then applied in incremental steps, to which there corresponds a piecewise linear response with material constants depending upon loading history and phenomenological characteristics of the material. This approach is appropriate for investigations of structural response to earthquake or wind loads in the nonlinear range [219,220].

For the analysis of structures or elements with strictly

geometric nonlinearities equivalent linearization procedures have been developed (see Refs. 217, 221, 222).

2.3 FUNDAMENTALS OF AEROELASTICITY

Aerodynamic forces induced by structural motions are referred to as self-excited. Aeroelasticity is the study of phenomena wherein the role of the self-excited forces is significant. Aeroelastic phenomena of interest in engineering applications include: lock-in, galloping, torsional divergence, and flutter.

2.3.1 Lock-In

This frequency at which vortices are shed in the wake of a rigid cylindrical body is equal to the Strouhal frequency, n_s, given by Eq. 2.1.7.

If the cylinder is elastically sprung, distributed pressures on the body associated with the vortices shed in its wake will cause the cylinder to oscillate. If the natural frequency of the cylinder differs significantly from the Strouhal frequency, the cylinder oscillations will be relatively small and will not affect the vortex shedding phenomenon, i.e., no aeroelastic interaction between the body motion and the fluid flow will be observed.

However, if the flow velocity is such that the Strouhal frequency is equal to the natural frequency of the body, n_1, i.e., if the flow velocity is

$$u_1 = n_1 D/S \tag{2.3.1}$$

where n_1 and D = natural frequency and diameter of body, respectively, and S = Strouhal number, resonant oscillations will occur. If the body is sufficiently flexible, the resulting oscillations will bring about significant changes in the wake flow, which in turn will affect the body motion. This process continues until a steady oscillatory state is reached.

It is observed that if the flow velocity changes slightly (up to a small percentage) from the value u_1 (Eq. 2.3.1), the vortex shedding frequency does not change in accordance with Eq. 2.1.7, but remains equal to n_1, i.e., the vortex shedding is controlled by the natural frequency of the body. This aeroelastic fluid-body interaction phenomenon is known as detuning or lock-in. Lift forces associated with locked-in oscillations may reach values one order of magnitude higher than those corresponding to the case of the rigid cylinder, and have caused numerous structural failures, notably of tall steel stacks.

Similar detuning phenomena occur in electrical engineering, where they can be described by certain nonlinear equations [223, p. 440, 224]. Similar nonlinear equations have also been employed in attempts to model lock-in, the earliest of which were proposed by

Bishop and Hassan [225] and Hartlen and Currie [226]. Reviews of research on vortex-excited oscillations are included in Refs. 202, 227, 228 and 229. Refs. 212 and 230 include information on the effect of turbulence in the oncoming flow upon such oscillations.

2.3.2 Galloping

Galloping is the term used to describe large amplitude oscillations occurring in a direction normal to the mean wind at frequencies much lower than those of vortex shedding from the same section. The oscillations can be several times larger than the across-wind dimensions of the section.

To explain intuitively the galloping phenomenon it is convenient to consider the case of a "D" -shaped section (Fig. 2.3.1). Assume that the body is initially at rest and that the flow velocity is normal to the face AB (Fig. 2.3.1 a). If, for any reason, the body experiences an incipient motion with velocity $\dot{y}$, the relative velocity of the flow with respect to the body will be equal to the resultant, V', of the velocities V and $\dot{y}$. (Fig. 2.3.1 b). The flow is now asymmetric, so that the flow speeds, and hence the pressures near points A and B, are unequal. Thus a lift force is generated along the y direction. This force accelerates the incipient motion of the body and thus has a destabilizing effect. A similar destabilizing effect occurs when, owing to the action of the restoring force, the body moves in the direction shown in Fig. 2.3.1 c.

Ice-coated power line cables are examples of bodies susceptible to galloping having cross-sections similar to the "D" -section of Fig. 2.3.1.

It can be demonstrated that the necessary condition for incipient galloping instability is

$$\left.\frac{dC_L(\alpha)}{d\alpha}\right|_{\alpha=0} + C_D(0) < 0 \qquad (2.3.2)$$

where α = angle of attack, [$\alpha = \tan^{-1}(\dot{y}/V)$ in Fig. 2.3.1 b], $C_L(\alpha)$ = mean lift coefficient corresponding to angle α, and $C_D(0)$ = mean drag coefficient for $\alpha = 0$ (Glauert-Den Hartog criterion [231]).

To determine whether a shape is susceptible to galloping instability, it is necessary to measure C_L for various angles of attack at and near $\alpha = 0$, and C_D for $\alpha = 0$. Eq. 2.3.2 must then be checked on the basis of values so measured.

Circular cylinders cannot gallop, since because of symmetry $dC_L(\alpha)/d\alpha = 0$ and therefore Eq. 2.3.2 cannot be satisfied. However, rectangular sections are susceptible to galloping, although the speeds required to cause it normally exceed by far the wind speeds of interest in building design.

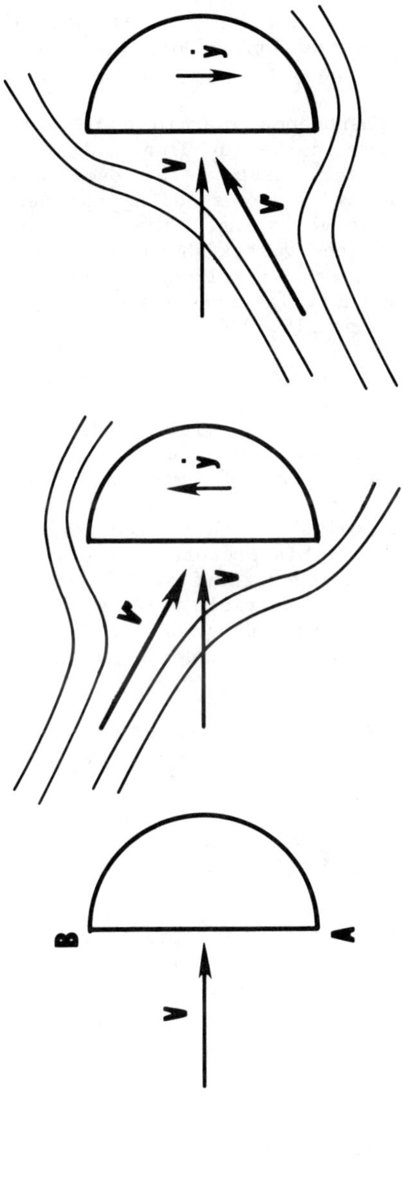

Figure 2.3.1 Galloping of "D" - shaped section.

Eq. 2.3.2 provides a criterion on the possibility that incipient galloping will occur. To study the evolution of galloping oscillations, information is required on the magnitudes of $C_L(\alpha)$ and $C_D(\alpha)$ for the entire range of α values occurring during the oscillations. Results based on experimental information used in conjunction with nonlinear models for galloping oscillations of prismatic structures are presented in Refs. 233 and 234. Refs. 234 and 235 include information on the effect of turbulence in the oncoming flow upon the galloping phenomenon.

Wake Galloping is the phenomenon wherein oscillations of a cylinder are induced by the flow in the turbulent wake of a second cylinder located upstream of the first. The oscillations are caused by the combined action of vortex shedding and aerodynamic instability phenomena similar to galloping. The frequency of the oscillations is low compared to the vortex-shedding frequencies of both the oscillating and the upstream cylinder. Wake galloping studies have been conducted notably in connection with power line cables grouped in bundles [202, p. 219, 236].

2.3.3 Torsional Divergence

Consider the section of Fig. 2.3.2 for which α = angle of attack, and k_α = torsional spring constant. To any particular flow velocity u, there corresponds an aerodynamic twisting moment. During the latter's action the angle α will increase. This will in turn result in an increased value of the aerodynamic twisting moment. For low flow velocities, this process eventually leads to a stable equilibrium position. However for flow velocities above a critical value the process becomes unstable. This phenomenon is analogous to the P-Δ effect for columns; the aerodynamic moment induced by the critical flow speed corresponds to the critical buckling load.

The critical speed is obtained from the relation expressing the balance of the internal structural moment, $k_\alpha \alpha$, and the external aerodynamic moment, i.e.,

$$k_\alpha \alpha = \frac{1}{2} \rho u^2 B^2 \left[C_{M_o} + \left.\frac{dC_M}{d\alpha}\right|_{\alpha = 0} \alpha \right] \qquad (2.3.3)$$

where ρ = air density, u = flow speed, B = width of section, C_{M_o} = value of C_M for $\alpha = 0$. It follows from Eq. 2.3.3 that α approaches infinity if u approaches the critical divergence velocity

$$u_c = \left[\frac{2k_\alpha}{\rho B^2 \left.\frac{dC_M}{d\alpha}\right|_{\alpha = 0}}\right]^{1/2} \qquad (2.3.4)$$

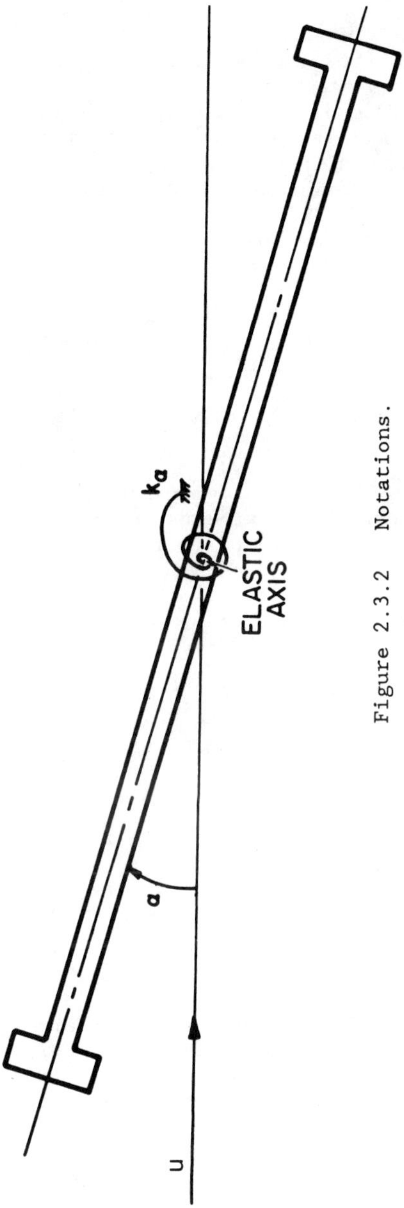

Figure 2.3.2 Notations.

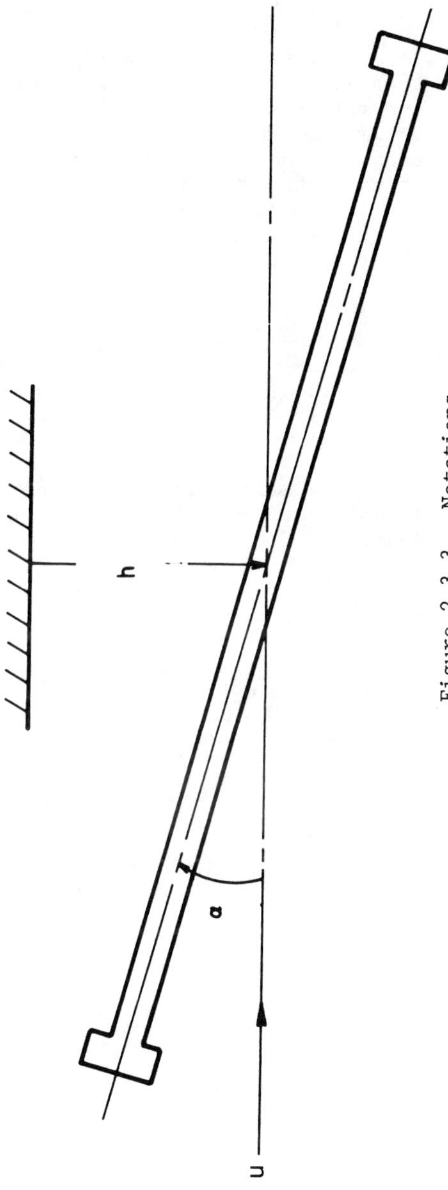

Figure 2.3.3 Notations.

2.3.4 Flutter

Flutter is an aeroelastic instability typical of structures such as airfoils or bridge decks that can experience both translation and torsional displacements [Fig. 2.3.3]. Let these displacements be denoted by h and α, respectively, and let m and I denote the mass and the moment of inertia per unit length of span, respectively. If the natural frequencies and the damping ratios in the translational and torsional modes are n_h, n_α, and ζ_h, ζ_a, respectively, the equations of motion of a bridge deck section with a span length equal to unity are:

$$\ddot{h} + 2(2\pi n_h)\, \zeta_h\, \dot{h} + (2\pi n_h)^2\, h = L_h/m \qquad (2.3.4\ a)$$

$$\ddot{\alpha} + 2(2\pi n_\alpha)\, \zeta_\alpha \dot{\alpha} + (2\pi n_\alpha)^2\, \alpha = M_\alpha/I \qquad (2.3.4\ b)$$

The aerodynamic lift and moment, L_h and M_α, are nonlinear functions of the displacements h and α and their time derivatives. However, the incipient flutter condition, which separates the stable and unstable regimes, may be treated as having only small amplitude. Therefore, for the study of flutter instability boundaries, L_h and M_α may be written as follows [237]:

$$L_h = \frac{1}{2}\, \rho\, u^2(2B)\ [K\ H_1^*(k)\ \frac{\dot{h}}{u} + KH_2^*(K)\ \frac{B\dot{\alpha}}{u} + K^2\ H_3^*\ (K)\alpha] \qquad (2.3.5\ a)$$

$$M_\alpha = \frac{1}{2}\, \rho\, u^2(2B^2)\ [KA_1^*(K)\ \frac{\dot{h}}{u} + KA_2^*(K)\ \frac{B\dot{\alpha}}{u} + K^2A_3^*(K)\alpha] \qquad (2.3.5\ b)$$

where K = reduced frequency is defined as

$$K = \frac{2\pi Bn}{u} \qquad (2.3.6)$$

n = frequency of oscillation of the airfoil or bridge deck, ρ = air density, u = flow speed, B = section width. The coefficients H_i, A_i (i=1,2,3) characterize the lift forces and the aerodynamic moments measured when the body is oscillating. Note that the typical term in Eqs. 2.3.5 follows the classical pattern of expressions for aerodynamic lift force, such as

$$L = \frac{1}{2}\, \rho u^2\ (2B)\ C_L \simeq \frac{1}{2}\, \rho u^2(2B)\ \frac{dC_L}{d\alpha}\ \alpha$$

for small angle of attack α.

Sample experimental values of the aerodynamic coefficients H_i^* and A_i^* for an airfoil, a partly streamlined box, and an open-truss deck are shown in Fig. 2.3.4. For purposes of comparison, Fig. 2.3.4 also shows these coefficients for an airfoil. If the aerodynamic coefficients H_i^* and A_i^* are determined experimentally, Eqs. 2.3.4 in conjunction with Eqs. 2.3.5 may be used to calculate the values of K and n at which the oscillations start to diverge. Let these values be denoted by K_c and n_c, respectively. The flutter (or critical) velocity is then

$$u_c = \frac{2\pi B n_c}{K_c} \qquad (2.3.7)$$

For certain types of bridges the values of the aerodynamic coefficients are such that the vertical and torsional motions are uncoupled, and single-degree-of-freedom flutter in torsion occurs. If, in addition, the coefficient A_3^* corresponding to the critical value of K is small, the torsional behavior of the deck at flutter is described by the equation of motion of a free oscillator with moment of inertia I, and natural frequency n_α. The critical velocity can be then determined from the condition that the coefficient of the term in $\dot{\alpha}$ of Eq. 2.3.4 b (combined with Eq. 2.3.5 b) be equal to zero. This yields

$$A_2^* \simeq \frac{2\ I\zeta_\alpha}{\rho B^4} \qquad (2.3.8)$$

Let the value of the reduced velocity for which Eq. 2.3.8 holds be denoted by K_c. The critical velocity is then given by Eq. 2.3.7 in which n_c is replaced by n_α.

Experimental studies on the effect of turbulence upon the onset of flutter are reported in Ref. 238. In the cases covered by these studies the critical velocity increases as a result of the presence of turbulence in the oncoming flow. A tentative explanation of this result is that the turbulence tends to reduce the coherence of the vortex flow in the structure's wake and thereby renders the fluid-body interaction mechanism less efficient than in the case where the oncoming flow is smooth. Note on the other hand that according to a theoretical approach proposed in Ref. 239, the turbulence should, on the contrary, bring about an increased susceptibility to flutter instability. No experimental corroboration of the theory of Ref. 239 appears to be available to date.

2.4 FUNDAMENTALS OF WIND TUNNEL TESTING

It has been shown previously that aerodynamic forces, including self-excited forces, may be significantly affected by the structure of the oncoming flow, i.e., the variation of mean speed with height, and features of the flow turbulence. For this reason, modern wind tunnel testing is usually conducted at facilities which attempt to simulate atmospheric flows.

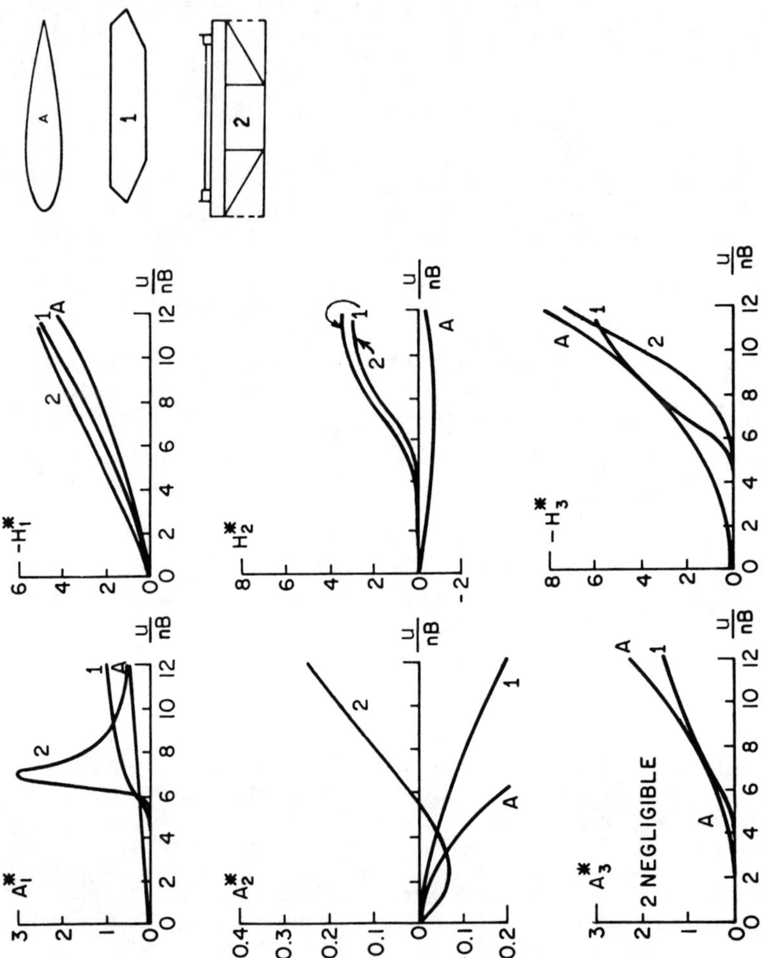

Figure 2.3.4 Aerodynamic coefficients for airfoil, partly streamlined box, and open-truss deck.

From Wind Effects on Structures, Second Edition, by Emil Simiu and Robert H. Scanlan. Copyright 1986 by John Wiley and Sons, Inc. Reprinted by permission of John Wiley and Sons, Inc.

This section will present basic information on wind tunnel testing as applied to structural engineering projects. More detailed information on this topic will be included in a separate report to be issued, in coordination with the Committee on Wind Effects, by the Committee on Aerodynamics, Aerospace Division, ASCE. In addition, information on the wind tunnel testing of various types of structure is presented as needed in subsequent sections of this report.

2.4.1 Similarity Requirements

It has been shown previously that aerodynamic forces, including self-excited forces, may be significantly affected by the structure of the oncoming flow, i.e., the variation of mean speed with height, and features of the flow turbulence. For this reason, modern wind tunnel testing is usually conducted at facilities which attempt to simulate atmospheric flows.

Similarity requirements pertaining to wind tunnel testing may be divided into three categories:

Geometric. These ensure similarity with respect to a characteristic nondimensionalized length.

Kinematic. These ensure similarity with respect to a characteristic nondimensionalized time, $\zeta = tu/D$ (t = characteristic time, u = characteristic flow speed, D = characteristic body dimension). For example, to the extent that kinematic similarity criteria are satisfied, the ratio of a nondimensionalized characteristic frequency of the fluctuating flow to a nondimensionalized natural frequency of vibration of an elastic body will be the same in the wind tunnel and in the atmosphere.

Dynamic. These ensure that the ratios between various forces are the same in the laboratory and at full-scale. For example, to reproduce correctly the ratio between fluid inertia forces to fluid viscous forces, the Reynolds number, Re, must be the same. It can also be shown that to reproduce correctly the ratios of fluid inertial forces to body inertial forces, elastic forces, and gravitational forces, the ratios σ/ρ, $E/\rho u^2$, and u^2/gD (the Froude number), respectively, must be the same in the laboratory and at full-scale, where σ = material density, ρ = air density, E = elastic modulus of the material, u = characteristic flow speed, g = acceleration of gravity, and D = characteristic dimension of body.

Certain similarity requirements cannot in practice be satisfied. In particular, in air tunnels the Reynolds number similarity requirement is inevitably violated. However, there is some evidence that failure to satisfy this requirement does not compromise the reliability of wind tunnel tests, both for structures with sharp edges and structures with curved shapes. Indeed, flows around the former are widely believed to be Reynolds number independent (see section 2.1.2.4), while flows around models of bodies with curved shapes can be rendered to some extent similar to full-scale flows provided that the body surfaces are treated appropriately (see section 2.1.3.1, and Refs. 202, 203, 206, and 207) and that aeroelastic effects are negligible [244].

Similarity requirements pertaining to the oncoming flow may be expressed as follows:

1. The ratios $u(z)/u(z_1)$ should be the same in the laboratory and at full scale for any z (u = mean speed, z = arbitrary elevation within the range of interest in the test, and z_1 = reference elevation).

2. All joint probability distributions of the various fluctuating velocity components should be the same in the laboratory and at full-scale.

Clearly, the application of the second requirement would be impractical. Instead, requirements used in wind tunnel practice include similarity with respect to: turbulence intensities; integral scales of turbulence; and spectra and cross-spectra of turbulent velocity fluctuations (see section 1.2.2).

Note that similarity with respect to spectra and cross-spectra automatically implies similarity with respect to both turbulence intensities and integral scales of turbulence. However, different spectral curve shapes will in general produce dissimilar integral scales, even if the areas under the respective spectral curves, and hence the r.m.s. of the turbulence fluctuations, are correctly scaled.

It appears that in certain applications primary attention needs to be given to the correct simulation of the turbulence intensity, the role of the turbulence scale being secondary [240]. However, in most applications, particularly those where the total fluctuating load acting over a given area is sought, it is important to simulate turbulence scales as correctly as possible. For example, assume that the objective of a test is to measure the total fluctuating load on an area A, and that the integral scales in the wind tunnel flow are half as large as the values obtained by applying the geometric scale to the integral scales of turbulence in the natural wind. It follows from the scaling requirement of section 1.2.2.1 that the wind tunnel measurements would correspond to the loading on an area four times as large as the area A. The wind tunnel measurements could thus underestimate the total fluctuating load acting on area A (see Eq. 2.1.16) [206].

2.4.2 Wind Tunnel Types

Wind tunnels used for structural engineering purposes may be divided into four basic categories:

- Aeronautical tunnels, in which the flow is uniform and smooth. Many of the aerodynamic data currently used in structural engineering practice were obtained in such tunnels. As indicated previously, tests in uniform smooth flow are currently seldom used for structural engineering purposes. Nevertheless, in situations where the critical wind loading may correspond to flows with relatively low wind speeds that could be relatively turbulence-free, it is appropriate to investigate the behavior of the structure in smooth flow (see, e.g., the investigation of wind loads on a pipe bridge reported in Ref. 241).

- Long tunnels, in which the atmospheric flows are simulated by a thick boundary layer that develops naturally over a rough floor.

- Tunnels with passive devices, in which a thick boundary layer is generated by grids, fences, or spires placed at the test section entrance. The flow is then allowed to pass over a fetch of roughness elements.

- Tunnels with active devices, such as jets or machine-driven shutters and flaps. In jet tunnels it is possible, within certain limits, to vary the mean velocity profile and the flow turbulence independently of each other [242]. Jet tunnels have been used primarily to study the effects of various uncertainties in the laboratory modeling of atmospheric flows [243].

Wind tunnel testing practice is currently not standardized, and discrepancies appear to occur, at least occasionally, among test results obtained in different wind tunnels or between results obtained from models and from prototypes. It is generally believed that as long as certain minimal performace criteria for wind tunnel testing are satisfied, such disprepancies and differences between laboratory and full-scale results can be kept within acceptable limits. The establishment of such performance criteria is an important task that remains to be carried out. Also needed by both structural designers and structural reliability specialists is information on the magnitude of errors in the estimation of wind effects inherent in the various wind tunnel techniques currently in use.

REFERENCES

201. Wooton, L., and Scruton, C., "Aerodynamic Stability", in The Modern Design of Wind-Sensitive Structures, Construction Industry Research and Information Association, London, U.K., 1971.

202. Simiu, E., and Scanlan, R. H., Wind Effects on Structures, Second Edition, John Wiley and Sons, New York, 1986.

203. Aérodynamique, Sciences du Bâtiment Volume II, Centre Scientifique et Technique du Bâtiment, Nantes, France, 1980.

204. Wind Forces on Structures, Transactions, ASCE, Vol. 126, Part II, 1961, pp. 1124-1198.

205. Laneville, A., Gartshore, I. S., and Parkinson, G. V., "An Explanation of Some Effects of Turbulence on Bluff Bodies," Proceedings, Fourth International Conference on Wind Effects on Buildings and Structures, Cambridge University Press, 1976.

206. Simiu, E., "Modern Developments in Wind Engineering," part 2, Engineering Structures, Vol. 3, October 1981, pp. 252-248.

207. Scruton, C., and Rogers, E. W. E., "Steady and Unsteady Wind Loading of Buildings and Structures," Philosophical Transactions of the Royal Society, London, Vol. A269, 1971, pp. 353-383.

208. Baines, W. D., "Effects of Velocity Distribution on Wind Loads and Flow Patterns on Buildings", in Proceedings of the Symposium on Wind Effects on Buildings and Structures, Vol. 1, National Physical Laboratory, Teddington, U.K., 1963, pp. 197-223.

209. Jensen M., and Frank, N., Model-Scale Tests in Turbulent Wind, Danish Technical Press, Copenhagen, 1065.

210. Bearman, P. W., "Turbulence Effects on Bluff Body Mean Flow", Proceedings, Third U.S. National Conference, Wind Engineering Research, University of Florida, Gainesville, 1978.

211. Vickery, B. J., and Clark, A. W., "Lift or Across-Wind Response of Tapered Stacks", Journal of the Structural Division, ASCE, Vol. 98, No. STI, Jan. 1972, pp. 1-20.

212. Griffin, O.M., Vortex Shedding from Stationary and Vibrating Bluff Bodies in a Shear Flow, NRL Memorandum Report 4287, Naval Research Laboratory, Washington, D.C., August 1980.

213. Whitbread, R. E., and Scruton, C., An Investigation of the Aerodynamic Stability of a Model of the Proposed Tower Blocks for the World Trade Center, New York, National Physical Laboratory Aero Report 1165, July 1965.

214. Hurty, W. C., and Rubinstein, M. F., Dynamics of Structures, Prentice Hall, Englewood Cliffs, N.J., 1964.

215. Haviland, R., A Study of the Uncertainties in the Fundamental Translational Periods and Damping Values for Real Buildings (Supervised by J. M. Biggs and E. H. Vanmarke), Publication No. R76-12, Order No. 531, Department of Civil Engineering, Massachusetts Institute of Technology, Feb. 1976.

216. Shinozuka, M., "Simulation of Multivariate and Multidimensional Random Process", Journal of the Acoustical Society of America, Vol. 49, No. 1, part 2, Jan. 1971, pp. 357-368.

217. Sinha, A. K., On Digital Simulation of Multicorrelated Random Process and Its Application, Department of Engineering Science and Mechanics, Virginia Polytechnic Institute and State University, July 1973.

218. Reed, D. A., "Time Series Analysis of Wind Loadings on Hyperbolic Cooling Towers," Ph.D. thesis, Dept. of Civil Engineering, Princeton University, Princeton, N.J., June, 1980.

219. Kanaan, A., Powell, G. H., "General Purpose Computer Program for Inelastic Dynamic Response of Plane Structures", EERC 73-6, University of California, Berkeley, Dec. 1973.

220. Dockholt, J. J., "INELAS2D" and "INELASTIER", Inelastic Analysis of Tier Buildings, TR158, Dept. of Civil Engineering, Stanford University, May 1972.

221. Yang, I., Stationary Random Response of Multidegree of Freedom Systems, Ph.D. thesis, California Institute of Technology, March 1970.

222. Carghey, T. K., "Equivalent Linearization Techniques," Journal of the Acoustical Society of America, Vol. 35, 1963, pp. 1706-1711.

223. Minorsky, N., Nonlinear Oscillations, Van Nostrand Company, Inc., Princeton, 1962.

224. Nayfeh, A. H., and Mook, D. T., Nonlinear Oscillations, John Wiley and Sons, New York, 1979.

225. Bishop, R. E. D., and Hassan, A. Y., "The Lift and Drag Forces on an Oscillating Cylinder", Proceedings of the Royal Society, Series A, Vol. 277, 1964, pp. 32-75.

226. Hartlen, R. T., and Currie, I. G., "Lift Oscillator Model of Vortex-Induced Vibration", Journal of the Engineering Mechanics Division, ASCE, Oct. 1979, pp. 577-591.

227. Dowell, E. H., "Nonlinear Elasticity in New Approaches to Nonlinear Problems in Dynamics, P. J. Holmes (ed.), SIAM, Philadelphia, 1980.

228. Sarpakaya, T., "Vortex-Induced Oscillations, A Selective Review", Transactions, ASME, Journal of Applied Mechanics, Vol. 46, pp. 241-258, 1979.

229. Blevins, R. D., Flow-Induced Vibration, Van Nostrand Reinhold, New York, 1977.

230. Howell, J. F., and Novak, M., "Vortex Shedding from Circular Cylinders in Turbulent Flow", Proceedings, Fifth International Conference on Wind Engineering, Vol. 1, Colorado State University, Fort Collins, 1979.

231. Den Hartog, J. P., Mechanical Vibrations, Fourth ed., McGraw-Hill, New York, 1956.

232. Scanlan, R. H., and Wardlaw, R. L., "Reduction of Flow-Induced Structural Vibrations," in Isolation of Mechanical Vibration, Impact, and Noise, AMD Vol. 1, Sec. 2, ASME, New York, 1973, pp. 35-63.

233. Parkinson, G. V., and Smith, J. D., "The Square Prism as an Aeroelastic Nonlinear Oscillator," Quarterly Journal of Mechanics and Applied Mathematics, Vol. 17, Part 2, 1964, pp. 225-239.

234. Novak, M., "Galloping Oscillation of Prismatic Structures," Journal of the Engineering Mechanics Division, ASCE, Vol. 98, No. EM2, Feb. 1972, pp. 27-46.

235. Kwok, K. C. S., and Melbourne, W. H., "Freestream Turbulence Effects on Galloping", Journal of the Engineering Mechanics Division, ASCE, Vol. 106, No. 106, No. EM2, Apr. 1980, pp. 273-288.

236. Wardlaw, R. L., Cooper, K. R., and Scanlan, R. H., "Observations on the Problem of Subspan Oscillation of Bundled Power Conductors", DME/NAE Quarterly Bulletin No. 1973(1), National Research Council, Ottawa, Canada, 1973, pp. 1-20.

237. Scanlan, R. H., and Tomko, J. J., "Airfoil and Bridge Deck Flutter Deviatives", Journal of the Engineering Mechanics Division, ASCE, Vol. 97, No. EM6, Dec. 1971, pp. 1717-1737.

238. Scanlan, R. H., and Lin, W. H., "Effect of Turbulence on Bridge Flutter Derivatives," Journal of the Engineering Mechanics Division, ASCE, Vol. 104, 1978, p. 719.

239. Lin, Y. K., Stochastic Analysis of Bridge Motions in Large Scale Turbulent Winds," Proceedings, Fifth International Conference on Wind Engineering, Fort Collins, Colorado, July 1979.

240. Marshall, R. D., Reinhold, T. A., and Tieleman, H. W., "Wind Pressures on Single-Family Dwellings", Proceedings ASCE/EMD Specialty Conference, University of California, Los Angeles, 1976.

241. Gartshore, I. S., Khanna, J., Laccinole, S., "The Effectiveness of Vorter Spoilers on a Circular Cylinder in Smooth and Turbulent Flow", Proceedings, Fifth International Conference on Wind Engineering, Fort Collins, Colorado, July 1979.

242. Hunt, J. C. R., and Fernholz, H., "Wind Tunnel Simulation of the Atmospheric Boundary Layer: A Report on Euromech 50," Journal of Fluid Mechanics, Vol. 70, part 3, Aug. 1975, pp. 543-559.

243. Corke, T. C., and Nagib, H. M., "Wind Loads on a Building Model in a Family of Surface Layers," Journal of Industrial Aerodynamics, Vol. 5, No. 182, October 1979, pp. 159-177.

244. Vickery, B. J., "The Aeroelastic Modeling of Chimneys and Towers, Proceedings, Wind Tunnel Modeling for Civil Engineering Applications, Gaithersburg, MD, April 1983, Cambridge University Press, Cambridge - New York, 1982.

3. LOW BUILDINGS

3.1.1 General Background

Much of the early research into wind loading problems was concerned with low buildings [301 to 309]. This early work included a number of full-scale measurements [306, 307, 308] as well as a remarkable set of wind tunnel tests conducted in boundary layer flow [309]. However, it was only in the 1970's that modern simulation and measurement techniques [310] as well as conceptual frameworks accounting for the important role of pressure fluctuations were applied to the low building problem in a systematic way. These recent efforts have led to an improved knowledge of the wind loads acting on a number of types of structures and will be reviewed in some detail herein.

The objective of this report is to present the current state-of-the-art in various areas of wind engineering and not to suggest or adopt provisions of any particular standard or code. In keeping with the evolutionary character of standards and codes, the updating of such provisions usually reflects compromises between commonly accepted practice and state-of-the-art results. This explains why pressure coefficients in this report are in some cases different in format and/or magnitude from corresponding values found in various codes and standards.

3.1.2 Experimental Aspects of Wind Load Estimation

Wind loads on low buildings can be estimated from full scale tests or by modeling in a wind tunnel. Low buildings have high natural frequencies and are not expected to vibrate under wind action. For this reason rigid models may be employed in the tests. Because atmospheric turbulence and wind speed gradients are strongest near the ground, the wind loads on low buildings exhibit strong fluctuations, the effects of which must be evaluated with care. Both external and internal unsteady pressures are important.

In wind tunnel testing of low buildings, geometric scales are small, i.e., they range from 1:200 to 1:2000 [311]. This creates problems related to instrumentation and the modeling of architectural details such as eaves and parapets. Two methodologies have been applied with a view to solving these problems. The first methodology attempts the simulation of the lower region of the atmospheric surface layer and employs larger model scales, 1:50 say [312, 313, 314]. By comparing full-scale and wind tunnel results, reference 314 suggests that proper simulation of the fluctuating wind pressures can be achieved by closely simulating the turbulence intensity and developing a turbulence integral scale at least as large as the largest model dimension; however, this suggestion has not been firmly established. The second methodology attempts to evaluate the errors in the wind pressures caused by a relaxation of the geometric scale requirement. These errors were shown to be acceptable in most cases if the model scale is exaggerated by a factor of up to two [315, 316].

Information required for design purposes generally consists of short duration peak loads acting over the tributary area of interest. For loads over larger areas (as opposed to pressures at local points), a simple economical method of measuring both mean and peak loads, referred to as pneumatic averaging, has been developed recently [317]. This method has been applied to the estimation of loads on building parts [318] and, in combination with computer weighting, on primary structural systems [319, 320]. Area loads have also been determined by conventional techniques, i.e. electrically summing the outputs of several pressure transducers or load cells [321, 322] but this method is expensive because the accurate evaluation of wind loading requires a large number of pressure transducers or costly load measurement units.

The duration of the peak load action is a significant factor affecting structural behavior. For example, the collapse load for steel frames increases substantially if the time of application of the load decreases [323]. Pressure measurement systems used in most wind tunnels suitable for building aerodynamics testing respond to pressure fluctuations on the model of the order of 100 Hz with negligible attenuation or distortion. For typical length and velocity scales, this corresponds to averaging times of roughly a second full scale. This appears to be a reasonable averaging time for most design purposes. Peak pressures coefficients based on shorter durations are higher than those based on one-second intervals. This was shown, for example, by measurements of peak pressures based on 1/32 sec duration of gust [324]. Further research is needed to evaluate the possible need for measuring peak loads averaged over intervals shorter than a second and to develop suitable laboratory techniques for such measurements.

3.1.3 Experimental Results

The major findings of experimental studies carried out either at full-scale or in wind tunnels simulations atmospheric flows are summarized below. Design data consistent with these results are given in section 3.1.4.

3.1.3.1 Influence of Fluctuation Loads

The fluctuating part of the loading is usually significantly larger than its mean component. This is particularly true of local pressures, and is true to a lesser degree of distributed load effects. An example time history of a local pressure is shown in Fig. 1.

3.1.3.2 Influence of Terrain Roughness

In general, the rougher the terrain, the larger the magnitude of the pressure coefficients but trends are not always consistent. However, the peak pressures associated with a particular storm decrease somewhat as terrain roughness increases, as shown in some detail in reference 325. This alleviation is not currently recognized in most standards and codes of practice for the design of low

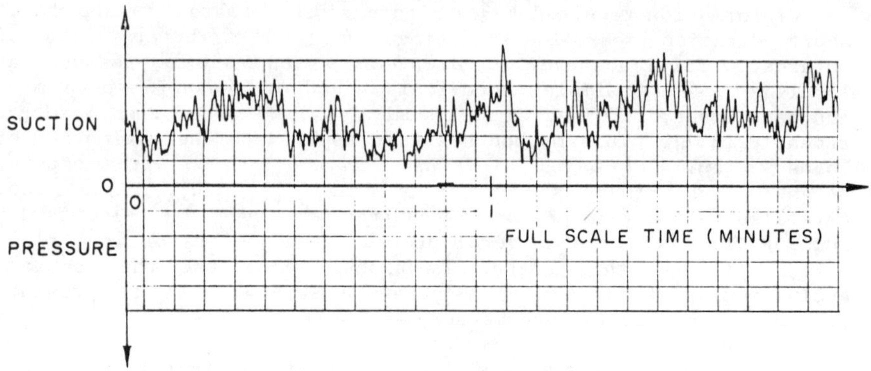

Fig. 1 Typical Trace of Suction Acting on a Point at the Edge of a 1:12 Slope Roof.

buildings. Rougher terrain also generally increases the ratio between the fluctuating and the mean component of the load.

3.1.3.3 Influence of Wind Direction

The strong effect of wind direction on the magnitude and distribution of wind loads has been shown in almost all studies of wind effects on low buildings. Considering the fact that, generally, no detailed analysis of the wind climate is carried out for the selected location of a low building and that the designers of most prefabricated buildings do not know in advance the actual building orientation on site, it has been suggested that, for design, the worst loads that would occur regardless of wind direction should be considered. However, such design would obviously involve considerable conservatism. Detailed reliability analyses are required to establish design loads on a realistic probabilistic basis. A first step in this direction has been described in ref. 326, and somewhat reduced design pressure coefficients for low buildings have been adopted by the Supplement to the National Building Code of Canada [327]. A more rigorous approach based on the joint probability distribution of wind speeds and pressure coefficients including the directional dependence is described in reference 328 and a simple reliability-based procedure for taking wind direction effect into account is presented in reference 386.

3.1.3.4 Influence of Geometric Parameters

For buildings with small height-to-width ratios and length-to-width ratios from 1.0 to 3.0 the wind loads do not depend significantly on the building length [325, 329, 330]. Wind loads increase with the building height but it was found [325] that the dependence of pressure coefficients on height can be reduced considerably by referencing them to the velocity pressure at mean roof height (for low roof slopes the eave height may be substituted for mean roof height). The roof slope strongly affects the magnitude and distribution of pressure coefficients since this is the main factor determining the wind flow pattern above the roof. Reference 331 has examined pressure coefficients acting on low buildings with three different roof angles (0°, 22.5^o and 45°) and has concluded that the high suctions appearing at edges and corners of flat roofs decrease for the 22.5^o roof and disappear on the 45° roof. Reference 332 reports that for roof angles of 0° and 6° the pressure distribution is essentially continuous across the ridge; at 12° there is a marked change at the ridge and the pressure coefficients immediately behind the ridge are similar to those at the leading edge but fall away rapidly towards the trailing edge; and at 22°, the leeward roof pressure coefficients are roughly constant. Several standards, as for instance the 1977 issue of the Canadian specifications [333], fail to specify high suction loads for ridge regions of intermediate roof slopes (roof angle around 20°). Results on roof pressure coefficients may also be found in references 318, and 334 through 343.

3.1.3.5 Influence of Tributary Area - Area loads

The effective pressures acting on roof or wall panels can be considerably less than the very high instantaneous pressures encountered locally. This is particularly true near roof corners, ridges and wall edges. References 318 through 322 give detailed information about the load reduction applied on various tributary areas.

3.1.3.6 Influence of Architectural Features

Architectural features such as mullions, overhangs and parapets affect the wind flow in the building proximity and are able to modify the effective wind loads on low buildings. There is a very limited amount of information with respect to the influence of architectural features and much research is still required on the subject. Studies dealing with the effect of mullions on tall buildings are reported in references 344 and 345.

The effect of canopies (also called eaves or overhangs) on the wind loading of low buildings, as well as the wind loads on eaves, have been examined in studies reported in references 329, 336, 342 and 346. Local pressure measurements indicate severe loading of windward canopies. This is due to the deflected flow from the windward wall which gives rise to a pressure on the lower canopy surface, reinforcing the high suction of the upper canopy surface right after separation. It has been found however, that canopy corners are less severely loaded than the corresponding roof corners of buildings without canopies [346].

The effect of parapets on the magnitude and distribution of pressure coefficients on low buildings has been a controversial subject. Reference 347 concludes that parapets do not have any effect on the mean roof loading. In contrast, results of reference 335 show that, in the presence of parapets, local mean suction loads acting on roofs become worse whereas the general level of peaks remains unaltered. A recent study [348], however, has shown that buildings with ratios of parapet height to building height ranging between 0.125 and 0.250 experience a reduction of the local highest roof suctions and a moderate increase of the remaining roof suctions. Mean roof suctions are somewhat lower, with the exception of the roof corners for which the parapets induce very high mean suctions and strong correlation of the local loads.

Some information on the effect of balconies on roof mean pressure coefficients is given in reference 330. For square buildings roof pressure coefficients increase with the number of balconies, particularly in the corner regions. The opposite trend appears for rectangular buildings with large length to width ratios.

Finally, reference 349 describes a study of wind pressure distribution on flat roofs with different edge profiles. Both mean and peak pressures have been found to be lower when the edge of a flat roof is rounded. This effect occurs for all wind directions and is more pronounced the greater the rounding. These results, however, may be Reynolds number dependent.

3.1.3.7 Influence of Large Nearby Structures

Nearby structures generally tend to reduce the wind loads acting on low buildings but there are some cases of close proximity and/or large relative heights where loads are increased. Cases of adverse wind loading induced by adjacent buildings are described in references 350, 351, 352, 353 and 354, whereas no adverse effects have been reported in [332] and [355]. A simple explanation for these adverse effects is provided by Baines [356]. The pressure on the front face of a building decreases downwards due to decreasing velocity in the boundary layer. Consequently, the pressure gradient induces a downward flow which can result in substantial velocities (and pressures) at lower levels. The tall building in effect amplifies the wind speeds experienced by the lower building. The clearance between the tall building and the low building is a significant factor for these adverse effects. It appears that a clearance at least two to three times the tall building width is necessary to avoid critical interaction. More details about these effects can be found in ref. 111.

3.1.3.8 Influence of Internal Pressure

Recent experimental and analytical studies [320 and 357 through 359] have shown that internal pressures are time dependent with gust factors only slightly lower than those applied to external pressures. The internal pressures are also well correlated spatially, which implies that their time dependent nature should be considered for structural loads.

The largest internal pressures occur when the wind direction is perpendicular to the wall with dominant openings. When dominant openings are to the lee of the structure and the windward wall is closed, the internal pressures are generally negative (internal suctions) and not very sensitive to the size of the openings or to the general permeability of the building envelope. Inertial effects of the air moving in and out of the building have also been found significant if a sudden change in external pressure, such as that due to a window failure, occurs for large openings [360]. External pressure distributions appear unaffected by either the permeability or the openings on the building envelope when large flows are not permitted through the building [320, 361, 362].

One of the major problems encountered in the determination of internal pressures is associated with the permeability of buildings, i.e., what constitutes normal leakage or effective openings for nominally closed buildings. References 320, 363, 364, and 365 provide some data for air leakages through different types of buildings but climatic conditions and construction techniques in different countries or regions yield a variety of building permeability ratios.

The effect of internal pressures in cavities such as those under tiled roofs has been studied recently [366]. It was reported that an impermeable backing to an air-permeable cladding reduces the wind load on the cladding. Additional experimental results on this effect are presented in reference 367. Some analytical work has also been carried out for the prediction of internal pressures from the external pressure distribution and results compare well with the experimental data [368]. Discussion of design criteria for internal pressures can be found in reference 369.

3.1.4 Design Wind Loads for Low Buildings

Most standards and Codes of Practice suggest some variation of the following general form for the calculation of total (external-internal) wind pressure acting on low buildings.

$$P = q\, C_e C_g (C_{pe} - C_{pi}) \qquad (3.1.4.1)$$

where q is the dynamic velocity pressure ($q = 1/2\, \rho\, V^2$, ρ = air density and V = wind speed averaged over a specified time duration and measured at a specified height, usually at 10 m or at roof height), C_e is an exposure factor accounting for the variation of the wind speed with height and terrain exposure, (for example, if Eq. 1.2.5 is used, the National Building Code of Canada (NBC) suggests the expression given in Table 1 for the determination of C_e), C_g is the gust factor reflecting the gustiness of the wind and the fluctuating character of the pressures, and C_{pe} and C_{pi} are the external and internal pressure coefficients, respectively (shape factors) which are determined experimentally. The validity of the pressure coefficients measured in wind tunnel studies is usually checked against corresponding full-scale measurement procedures. Comparisons between full-scale and wind-tunnel data are described in references 329, 338 and 370 through 374.

TABLE 1 - EXPOSURE FACTORS SUGGESTED BY NBC

Terrain	Exposure Factor (C_e) at height z (m)
Open	$C_e = (\frac{z}{10})^{0.28}$, $C_e \geq 1.0$
Suburban	$C_e = 0.5\ (\frac{z}{12.7})^{0.50}$, $C_e \geq 0.5$
Urban	$C_e = 0.4\ (\frac{z}{30})^{0.72}$, $C_e \geq 0.4$

The American National Standard ANSI A58.1 [375] has adopted the equation

$$P = qG_hC_p - q_h\ (GC_{pi}) \qquad (3.1.4.2)$$

for the design pressure of the main wide-force resisting system and the equation

$$P = q_h\ (GC_p) - q_h\ (GC_{pi}) \qquad (3.1.4.3)$$

for the design of components and cladding of low buildings (of height $h \leq 60$ ft (18.3 m). Equations 3.1.4.2 and 3.1.4.3 are of the same form as Equation 3.1.4.1. (In Eqs. 3.1.4.2 and 3.1.4.3 the gust factor is denoted by G and the external pressure coefficient by C_p; the influence of exposure (C_e) has been incorporated into G_h and q in Eq. 3.1.4.2 and has been eliminated from Eq. 3.1.4.3 by assuming open terrain exposure for all environments). The dynamic velocity pressure at mean roof height, q_h, of Eq. 3.1.4.3 is given by the relationship $q_h = 0.00256\ K_h\ V_f^2$ in which K_h is a height factor tabulated in ref. 375, V_f is the fastest mile speed (mph) and q_h is given in psf. Finally, q of Eq. 3.1.4.2 equals q_h for all building surfaces except the windward wall for which it equals q_z at height z above ground.

Despite the fact that all standards and codes of practice use the general form of equation 3.1.4.1, there are significant differences as to the wind speeds employed for the determination of the dynamic velocity pressure q. Different return periods, averaging times and reference heights are used in various standards. It is well known, for instance, that the 25, 50 and 100-year return period fastest mile wind speed is used in the United States, whereas the mean hourly speed with 10,30 or 100-year return period is used in Canada and a 3-second gust speed with a 50-year return period is used in Great Britain. The pressure coefficients suggested by various standards for different types of buildings are adjusted accordingly. These adjustments should be taken into account when comparisons among various standards are attempted.

The majority of standards and codes of practice worldwide suggest pressure coefficients for low buildings based on experiments not simulating the natural wind characteristics. The National Building Code of Canada and the ANSI Standard are among the few which have introduced pressure coefficients for simple building forms based on experiments conducted in wind tunnels simulating the natural wind conditions.

Research studies aiming at the definition of wind loads acting on low buildings in a simple codified form have been carried out in various countries [Australia, Canada, France, Great Britain and U.S.A.]. Reference 331 (pp. 50-51) suggests values of mean and rms pressure coefficients for low buildings of medium roof angle (22.5^o) for wind perpendicular to the side wall. Two different roughnesses (power-law exponents of 0.13 and 0.20) and the effect of nearby environment have been considered. Peak pressure coefficients can be calculated by using suggested pressure probability functions [376]. Additional results including more roof angles and wind directions appear in Ref. 377. These data include peak pressure coefficients but they are based on a limited number of pressure tap measurements.

Recommended wind load design coefficients for mobile homes have been included in reference 321. Of particular interest is the fact that the effect of area load reduction has been measured and considered in the derivation of wind loads. References 341 and 342 suggest design peak pressure coefficients for single-story and elevated single-story (high-set)houses of roof angle equal to 10° with overhangs.

A model of peak pressure coefficients (C_gC_{pe} amd C_gC_{pi} of equation 3.1.4.1) for flat or gabled roof buildings with eave heights less than 60 ft. (18.3 m) and length to width ratios > 1 has been provided [326] as the result of the comparison and selection of a large amount of data [320, 325]. The model has been derived from data selected from experiments in a simulated open country exposure and it is slightly conservative if used in conjunction with any other terrain roughness. <u>Values suggested are referenced to the dynamic velocity pressure based on the hourly mean speed measured at the mean roof height</u> and are valid for all azimuths. The hourly mean speed may be calculated from the fastest mile wind speed following the procedure described in section 1.2.2.6. The model is summarized in Figs. 2 and 3 and in Table 2.

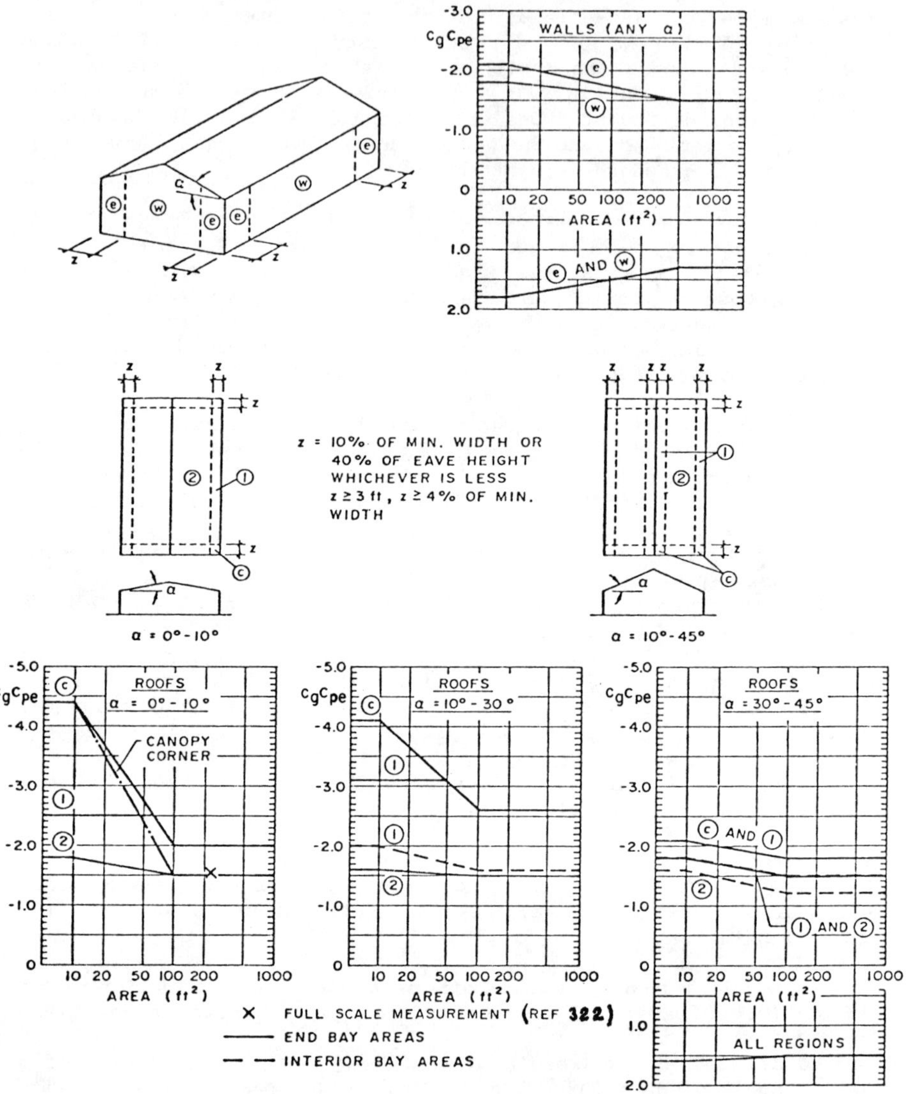

Figure 2. External Pressure Coefficients for Maximum Loads on Cladding and Secondary Structural Elements.

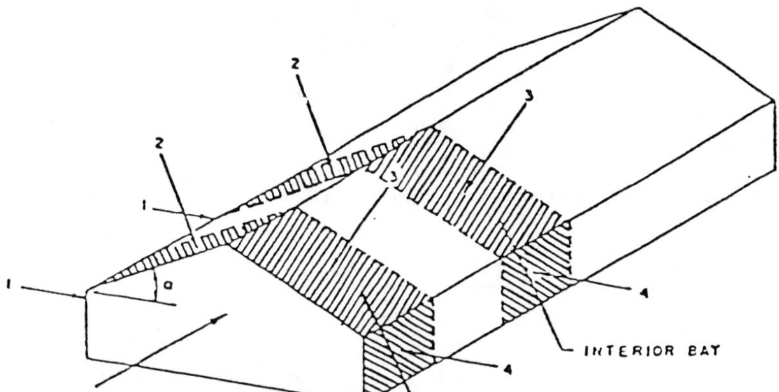

	End Bay Coefficients			C_gC_{pe}	Interior Bay Coefficients			
α	1	2	3	4	1	2	3	4
0° - 10°	1.15	-2.00	-1.00	-0.80	0.75	-1.30	-0.70	-0.55
10° - 30°	1.50	-2.00	-1.30	-1.20	1.00	-1.30	-0.90	-0.80
30° - 45°	-0.90 1.30	-2.00 0.50	-1.00 -1.00	-0.90 -0.90	-0.85 1.05	-1.30 0.40	-0.70 -0.80	-0.85 -0.70

These coefficients are only to be used in combination so as to develop appropriate loads for the design of the primary structural system.

Foundations: For the design of foundations, but exclusive of anchorages to the frame, only 70% of the effective loads is to be considered.

Figure 3. External Pressure Coefficients for Primary Structural Loads on Structural Systems.

TABLE 2 - INTERNAL PEAK PRESSURE COEFFICIENTS

	Max C_gC_{pi}	Min C_gC_{pi}
Openings Uniformly Distributed	0	- 0.3
All Other Cases	1.0	-0.3

Figure 2 shows external peak pressure coefficients associated with various tributary areas of structural interest. Different values are suggested for the various zones, as indicated. The end-bay areas have a minimum width 2z or 20 ft. (6 m) for buildings without frames. The net effect of applying the loading patterns appearing in Fig. 3 is to ensure that all the major wind actions found experimentally are enveloped. These wind actions include total uplift, horizontal thrust and bending moments at critical points. Table 2 shows the range of peak internal pressure coefficients (C_gC_{pi}). The value which, together with the external pressure coefficient, would give the most critical pressure should be used.

Design peak pressure coefficients (i.e., C_gC_{pe} values of Equation 3.1.4.1) for elevated single-story buildings high-set houses)are given in reference 341.

Codified mean pressure coefficients for low buildings with multiple roof configurations have been suggested in reference 378 following an extensive program of full-scale measurements. Figure 4 shows the types and geometrical characteristics of the experimental buildings studied and Fig. 5 shows the suggested values of external pressure coefficients. It should be mentioned here that these pressure coefficient values are appropriate for use with the British Code of Practice [379] and as such they cannot be compared directly with values suggested previously (e.g. Fig. 2) for, say, a single-span gabled house. To obtain design pressures, the coefficients of Fig. 5 should be multiplied by $1/2\,\rho\, V_s^2$, where $V_s = VS_1S_2S_3$ to yield external pressures. In accordance with the British Code, V is a 3-sec gust wind speed at 10 m height in open level country and S_1, S_2, S_3 are factors reflecting local topographic influences, surface roughness of the environment, gust duration appropriate to the size, the ridge height and the design life of the building. The 3-sec gust wind speed can be calculated from the fastest mile or hourly mean wind speed following the procedure described in section 1.2.2.6. It may be assumed S_1=1 except for exposed hills and valleys, where S_1 = 1.1. Tables 3 based on reference 379 gives values of the factor S_2. For the evaluation of the S_2 factor (Table 4), class A column is used for the design of cladding, glazing and roofing including their fixings and class B column for the design of main structural systems for buildings whose greatest

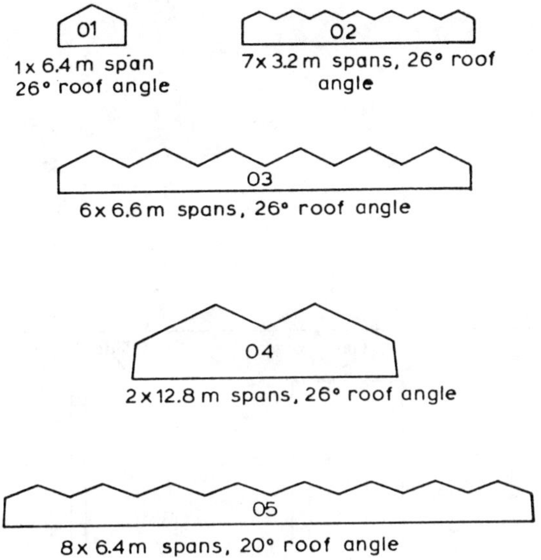

Details of **experimental** glasshouses

House ref.	No. of spans	Span width (m)	Overall width (m)	Length (m)	Eaves height (m)	Ridge height (m)	Roof angle (deg)	Eaves height/ span width ratio
01	1	6.4	6.4	21.3	2.4	3.9	26	0.37
02	7	3.2	22.4	63.0	2.4	3.1	26	0.73
03	6	6.6	39.7	79.6	2.4	4.0	26	0.36
04	2	12.8*	25.6	39.6	3.4	7.1	26	0.26
05	8	6.4	51.2	88.8	2.8	3.9	20	0.44

*Asymmetric spans.

Figure 4. Experimental Building Types and their Geometrical Characteristics. From D. A. Wells and R. P. Hoxey, "Measurements of Wind Loads on Full-Scale Glasshouses", Journal of Wind Engineering and Industrial Aerodynamics, Vol. 6, pp. 139-167, 1980.

End elevation | Plan | Side elevation

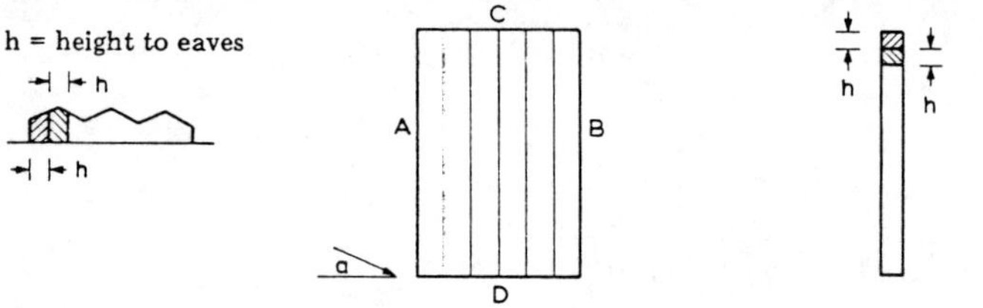

Wind angle α (deg)	C_{p_e} for surface				Local C_{p_e}	
	A	B	C	D	(hatched, left)	(hatched, right)
0	+0.6	−0.4	−0.3	−0.3	−1.2	−0.6
					on gable ends C,D	
90	−0.2	−0.2	+0.7	−0.4	−1.0	−0.5
					on side walls, A,B	

Figure 5a. Pressure Coefficients C_{pe} Wall of Single and Multi-Span Structures.

(a) Single-span

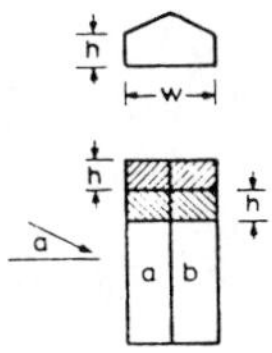

Wind angle α (deg)	Roof angle (deg)	C_{p_e} a	C_{p_e} b	Local C_{p_e} (hatched)	Local C_{p_e} (cross-hatched)
0	26	see Fig. 5 c	−0.6	−1.0	−0.8
				on roof slope b only	
	20	see Fig. 5 c	−0.6	−1.0	−0.8
				on roof slope b only	
90	20—26	−0.2	−0.2	−1.2	−0.5

(b) Multi-span

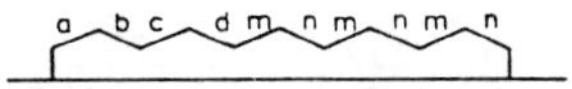

Wind angle α (deg)	Roof angle (deg)	h/w	C_{p_e} First spans a*	First spans b	Second span c	Second span d	Other spans m	Other spans n	Local C_{p_e}** (hatched)	Local C_{p_e}** (cross-hatched)
0	26	0.26	+0.3	−0.5	−0.5					
		0.36	−0.1	−1.2	−1.1	−0.6	−0.4	−0.5	—	—
		0.73	−0.8	−1.0	−0.7					
0	20	0.44	−0.85	−1.0	−0.8					
90	20—26	all	−0.2	−0.2	−0.2	−0.2	−0.2	−0.2	−1.4	−0.6

*As for single-span, i.e. pressure coefficients from Fig. 5c
**As designated for single-span and to be applied to all spans

Figure 5b. Pressure Coefficients C_{pe} for Pitch Roofs of Single and Multi-Span Structures.

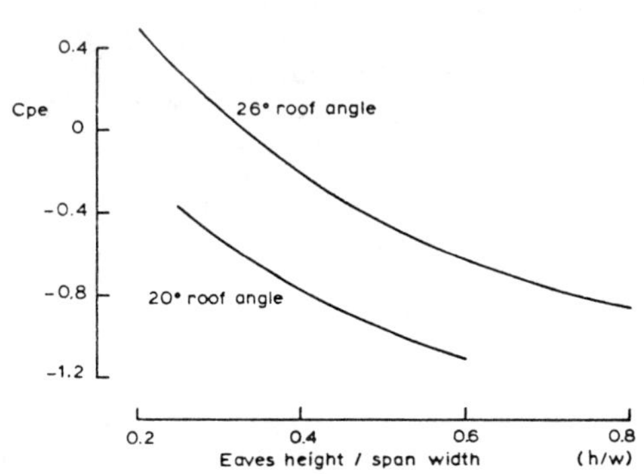

Figure 5c. Pressure Coefficients for First Roof Slope. From D. A. Wells and R. P. Hoxey, "Measurements of Wind Loads on Full-Scale Glasshouses", Journal of Wind Engineering and Industrial Aerodynamics, Vol. 6, pp. 139-167, 1980.

horizontal dimension does not exceed 50 m. For all other cases class C Column is to be used. It can be assumed that $S_3 = 1$ for the great majority of cases implying a 50-year return period wind. More details about the evaluation of design pressures can be found in reference 379.

TABLE 3 GROUND ROUGHNESS, BUILDING SIZE AND HEIGHT ABOVE GROUND FACTOR, S_2

H (meters)	(1) Open country with no obstructions			(2) Open country with scattered windbreaks			(3) Country w/many windbreaks: small towns; suburbs			(4) City centers		
	Class A	Class B	Class C	Class A	Class B	Class C	Class A	Class B	Class C	Class A	Class B	Class C
3 or less	0.83	0.78	0.73	0.72	0.67	0.63	0.64	0.60	0.55	0.56	0.52	0.47
5	0.88	0.83	0.78	0.79	0.74	0.70	0.70	0.65	0.60	0.60	0.55	0.50
10	1.00	0.95	0.90	0.93	0.88	0.83	0.78	0.74	0.69	0.67	0.62	0.58
15	1.03	0.99	0.94	1.00	0.95	0.91	1.00	0.95	0.91	0.74	0.69	0.64
20	1.06	0.99	0.94	1.03	0.98	0.94	0.88	0.83	0.85	0.79	0.75	0.70
30	1.09	1.05	1.00	1.07	1.03	0.98	1.01	0.97	0.92	0.90	0.85	0.79

Wind pressure coefficients presented or referred to so far were obtained from full-scale tests or from tests in wind tunnels in which an attempt was made to simulate natural winds. However, there is a large variety of building configurations which have not as yet been tested using this methodology. For many such configurations wind loading data obtained from wind tunnel tests in smooth flow (and, therefore, not representative of natural wind conditions) are reproduced from Refs. 305 and 380 in Refs. 381 and 111.

Numerical Example

Assume a gabled-type low building with dimensions as shown in Fig. 6. The building may have any length greater than its width. Assume that the basic wind speed (fastest-mile wind speed at 10 m above ground in open terrain) is $V_f = 115$ mph. Using the procedure described in 1.2.2.6 the hourly mean speed in open terrain at 10 m above ground is V(10) = 87.3mph (38.8m/s). The exposure factor, calculated from Table 1, is $C_e = (z/h)^{0.28} = 1.07$. Then, following Eq. 3.1.4.1, $C_e\ q = 1.07 \times 1/2 \times 1.24 \times 38.8^2 = 999\ N/m^2$, assuming $\rho = 1.24\ kg/m^3$. The external wind pressures (and/or suctions) for the design of cladding and secondary structural elements shown in Fig. 6 were calculated on the basis of the diagrams of Fig. 2 (peak

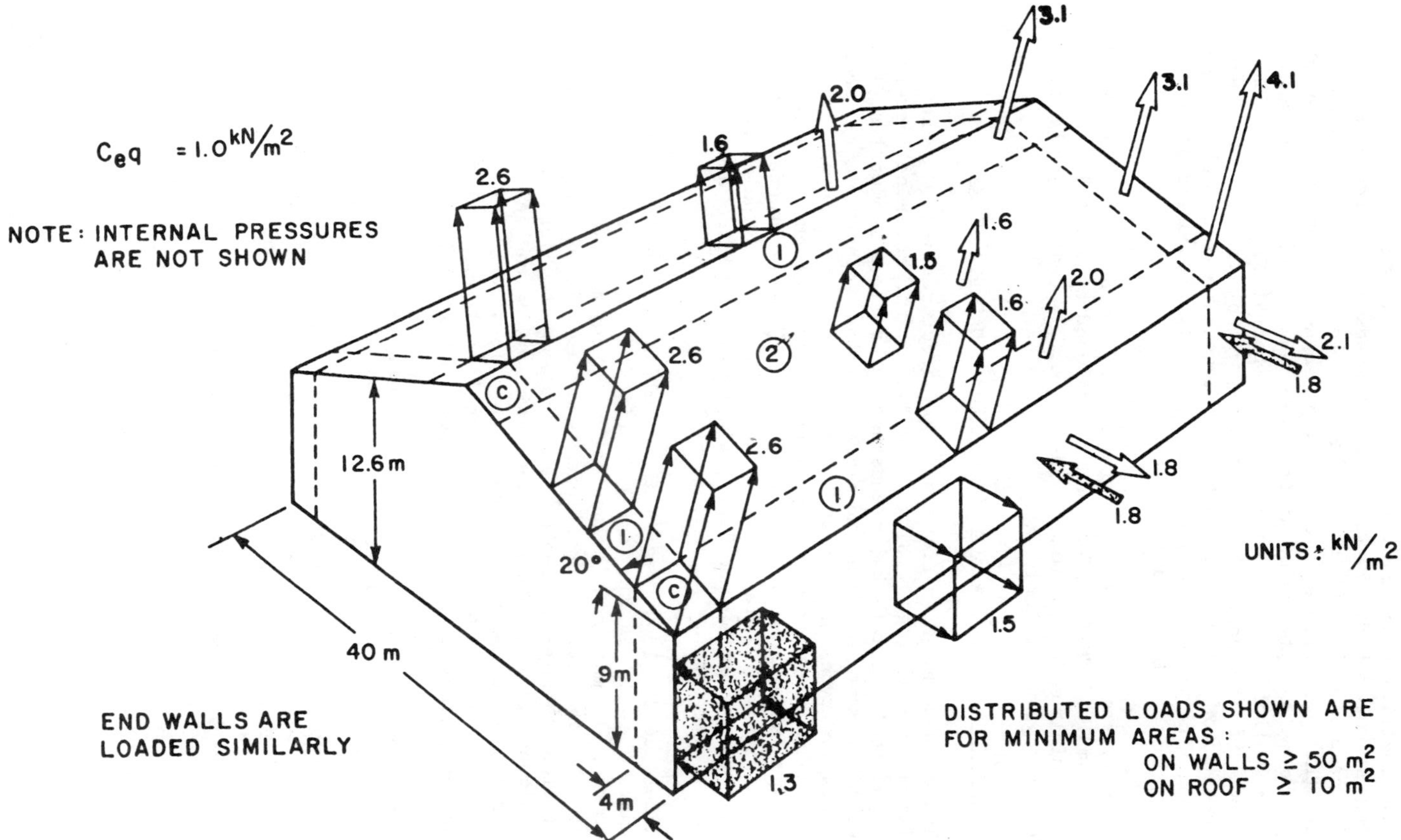

Figure 6. Example of Application of Pressures for the Design of Cladding and Secondary Structural Members of Various Tributary Areas for a Low Building.

pressure coefficients) and Eq. 3.1.4.1. For the design of the building structural system the primary loads suggested in Fig. 3 for the case α = 10°-30° should be used in combination so as to develop the appropriate total forces, bending moments, and so forth. Note that the calculations have been carried out assuming that the building is located in open terrain. As previously mentioned, if the approach illustrated in this example is used, this assumption can be employed regardless of the actual terrain conditions, the results so obtained being slightly conservative for buildings in suburban or urban exposures.

Finally note that internal pressures have not been considered in this example. Their effect can be included by using Eq. 3.1.4.1 with the minimum or maximum value of C_gC_{pi} (whichever is more unfavorable) given in Table 2.

3.1.5 Wind Loads for Low Structures Other than Buildings

There is a great variety of low structures which cannot be classified as buildings. Simple shelter roofs, solar collectors, small spherical water tanks, grandstands of various geometries, low shelter walls, traffic signs and signals, and noise barriers are some representative types of such structures. Wind load recommendations for those structures are mainly based on early wind tunnel experiments carried out under smooth flow conditions (see references 381 or 111). At present, there are very few wind loading studies carried out for such structures at full scale or in simulated atmospheric flow, e.g., ref. 382.

Reference 383 reports a detailed full-scale and boundary layer wind tunnel study on the determination of wind loads on solar collectors. The main conclusions of this study may be summarized as follows: (1) peak pressure coefficients are usually negative for those installations whose collector inclination is less than 60°; (2) Magnitudes of peak pressure coefficients for the end collectors are higher than those for the collectors located in the middle of the row; (3) shielding by other rows of collectors and/or roof parapets reduces the magnitudes of the peak pressure coefficients.

Detailed values of design pressure coefficients for solar collectors have been compiled and are given in reference 384.

Wind loads on traffic signals have been investigated by Marchman (385) in a wind tunnel study. All tests were performed on free-swinging lights, and the signal was allowed to assume its natural orientation to the flow. Wind pressure coefficients for signs and outdoor display structures can be obtained from reference 375.

REFERENCES

301. Stanton, T.E., "Report on the Measurement of the Pressure of the Wind on Structures, "Proceedings, The Institution of Civil Engineers, Paper No. 4513, 1924.

302. Bounkin, K. and Tcheremoukhin, A., "Wind Pressures on Roofs and Walls of Buildings," Trans., Central Aero-Hydrodynamical Institute, Moscow, 1928.

303. Irminger, J.O.V. and Nokkentved, C., "Wind Pressure on Buildings," First series, Danmarks Naturvidenskabelige Samfund, 1930.

304. Irminger, J.O.V. and Nokkentved, C., "Wind Pressure on Buildings," Second series, Danmarks Naturvidenskabelige Samfund, 1936.

305. Ackeret, J., "Der Winddruck auf Schornsteine mit Kreisquerschnitt," Schweizer Bauzeitung, Vol. 108, No. 3, 1936.

306. Bailey, A., "Wind Pressures on Buildings," Selected Engineering Paper NO. 139, The Institution of Civil Engineers, 1933.

307. Bailey, A. and Vincent, N.D.G "Wind Pressure on Buildings, Including Effects of Adjacent Buildings," Journal of Institution of Civil Engineers, 20, 1943.

308. Arnstein, K. and Klemperer, W., "Wind Pressures on the Akron Air-ship Dock," Journal of Aeronautical Sciences, Vol. 3, Jan., 1936, pp. 88-90.

309. Prandt, L. and Betz, A. "Ergebnisse der Aerodynamischen Versuch- sanstalt zu Gottingen," Verlag von H. Oldenbourg, MU chen and Berlin 1932 p. 128.

310. Surry, D. and Isyumov, N., "Model Studies of Wind Effects - A Perspective on the Problems of Experimental Technique and Instrumentation," Proceedings of the Sixth International Congress on Instrumentation in Aerospace Simulation Facilities, Ottawa, Canada, Sept. 22-24, 1975.

311. Hunt, J.C.R. and Fernholz, H., "Wind-tunnel Simulation of the Atmospheric Boundary Layer: A Report on Euromech 50," Journal of Fluid Mechanics, Vol. 70, part 3, 1975, pp. 543-559.

312. Cook, N., "On Simulating the Lower Third of the Urban Adiabatic Boundary Layer in a Wind Tunnel," Atmospheric Environment, 7, 1973, pp. 691-705.

313. Holmes, J.D., "Design and Performance of a Wind Tunnel for Modelling the Atmospheric Boundary Layer in Strong Winds," Wind Engineering Report 2/77, Dept. of Civil and Systems Engineering, James Cook University of North Queensland, Australia, May, 1977.

314. Tieleman, H.W.Reinhold, T.A. and Marshall, R.U., "On the Wind-Tunnel Simulation of the Atmospheric Surface Layer for the Study of Wind Loads on Low-Rise Buildings," Journal of Industrial Aerodynamics, Vol.3, 1978, pp. 21-38.

315. Stathopoulos, T., Surry, D. and Davenport, G., "Some General Characteristics of Turbulent Wind Effects on Low-Rise Structures,' Proceedings of 3rd Colloquium on Industrial Aerodynamics, Aachen, W. Germany, June 14-16, 1978.

316. Simiu, E., "Modern Developments in Wind Engineering. Part 2," Engineering Structures, 1981, Vol. 3, October, pp. 242-248.

317. Surry, D. and Stathopoulos, T., "An Experimental Approach to the Economical Measurement of Spatially-Averaged Wind Loads," Journal of Industrial Aerodynamics, Vol. 2, 1977/1978, pp. 385-39.

318. Stathopoulos, T., Surry, D. and Davenport, A.G., "Effective Wind Loads on Flat Roofs", J. of the Structural Division, ASCE, Vol. 107, No. ST2, Proc. Paper 16039, Febr. 1981, pp. 281-298.

319. Surry, D., Stathopoulos, T. and Davenport, A.G., "Simple Techniques for the Direct Measurement of Area Loads Induced by Wind," J. of Engineering Mechanics, ASCE, Aug. 1983.

320. Stathopoulos, T., Surry, D. and Davenport, A.G., "Internal Pressure Characteristics Due to Wind Action," Proceedings of the 5th International Conference on Wind Engineering, Ft. Collins, Colorado, July 8-14, 1979.

321 Marshall, R.U., "The Measurement of Wind Loads on a Full-Scale Mobile Home," National Bureau of Standards, Washington, D.C, NBSIR 77-1289, Sept., 1977.

322. Kim, S. I. and Mehta, K. C., "Wind Loads on Flat Roof Area Through Full-Scale Experiment," Institute for Disaster Research, Texas Tech University, Lubbock, Texas 79409, Sept., 1977.

323. Baker, J. F., Horne, M.R. and Heyman, J., "The Steel Skeleton," Vol. 2, Plastic Behavior and Design, pp. 361-364, Cambridge University Press, 1956.

324. Eaton, K. J. and Mayne, J. R., "The Measurement of Wind Pressures on Two-Story Houses at Aylesbury," CP 70/74, Building Research Establishment, Department of the Environment, July, 1974.

325. Davenport, A. G., Surry, D. and Stathopoulos, T. "Wind Loads on Low-Rise Buildings: Final Report of Phases I and II-Parts 1 and 2," BLWT - SS8-1977, The University of Western Ontario, London,k Canada, Nov. 1977.

326. Stathopoulos, T., Surry, D. and Davenport, A.G., "A Simplified Model of Wind Pressure Coefficients for Low-Rise Buildings," Proceedings of the 4th Colloquium on Industrial Aerodynamics, Aachen, West Germany, June 18-20, 1980.

327. The Supplement to the National Building Code of Canada 1980, Associate Committee on the National Building Coed, National Research Council of Canada, Ottawa, NRCC No. 17724.

328. Cook, N.J. and Mayne, J.R., "A Novel Working Approach to the Assessment of Wind Loads for Equivalent Static Design", Journal of Industrial Aerodynamics, Vol. 4, 1979, pp. 149-164.

329. Tieleman, H.W. and Reinhold, T. A., "Wind Tunnel Model Investigation for Basic Dwelling Geometries," Virginia Polytechnic Institute and State University Report VPI-E-76-8, May, 1976.

330. Kramer, C., Gerhardt, H.J. and Scherer, S., "Wind Pressure on Block-type Buildings," Proceedings of 3rd Colloquium on Industrial Aerodynamics, Aachen, W. Germany, June 14-16, 1978.

331. Barnaud, G. and Gandemer, J., "Détérmination en Soufflerie Simulant le Vent Naturel des Coefficients de Pression sur les Structures Basses," Section Aérodynamique de Constructions ADYM-12.74, Centre Scientifique et Technique du Bâtiment, Establissement de Nantes, France.

332. Vickery, B. J., "Wind Loads on Low-Rise Buildings," presented at D.R.C. Seminar, Darwin, Australia, March 30, 1976.

333. Commentaries on Part 4 of the National Buildings Code of Canada, Supplement No. 4, Associate Committee on the National Building Code, National Research Council of Canada, Ottawa, NRCC No. 15558.

334. Jensen, M. and Franck, N. "Model-Scale Tests in Turbulent Wind, Parts I and II," The Danish Technical Press, Copenhagen, 1965.

335. Davenport, A.G. and Surry, D., "Pressures on Low Rise Structures in Turbulent Wind," Proceedings of Canadian Structural Engineering Conference 1974 (Canadian Steel Industries Construction Council).

336. Dreher, K. J. and Cermak, J.E., "Wind Loads on a House Roof," Colorado State University, College of Engineering Report CER 72-73, KJD-JEC22, March, 1973.

337. Hellers, G. and Lundgren, S., "Wind Loads on Generally Shaped House Bodies - Model Tests," Report R22, 1974, National Swedish Building Research Summaries.

338. Marshall, R. D., "A Study of Wind Pressures on a Single-Family Dwelling in Model and Full Scale," Proceedings of the Invitation Symposium on Full Scale Measurements of Wind Effects on Tall Buildings and Other Structures, Boundary Layer Wind Tunnel Laboratory, The University of Western Ontario, Canada, June 23-29, 1974.

339. Marshall, R. D., "A Study of Wind Pressures on a Single-Story Houses," Proceedings of Second U.S. National Conference on Wind Engineering Research, Colorado State University, Fort Collins, Colorado, June 22-25, 1975.

340. Eaton, K.J., Mayne, J.R. and Cook, N.J., "Wind Loads on Low-Rise Buildings - Effects of Roof Geometry," CPI/76, Building Research Establishment, Department of the Environment, January 1976.

341. Holmes, J.D. and Best, R.J. "Wind Pressures on an Isolated High-Set House," James Cook University of North Queensland, Wind Engineering Report 1/78, Queensland, Australia, Jan., 1978.

342. Best, R. J. and Holmes, J.D., "Model Study of Wind Pressures on an Isolated Single-Story House," James Cook University of North Queensland, Wind Engineering Report 3/78, Queensland, Australia, Sept., 1978.

343. Holmes, J.D., "Wind Pressures on Houses with High-Pitched Roofs," James Cook University of North Queensland, Wind Engineering Report 4/81, Queensland, Australia, Nov., 1981.

344. Templin, J. T. and Cermak, J.E., "Wind Pressure on Buildings: Effect of Mullions," Colorado State University, College of Engineering Report CER76-77 JTT-JEC24, Fall, 1976.

345. Leadon, B.M. and Kounacki, M.L., "Pressures on Curtain Wall with External Mullions," Journal of Engineering Mechanics Division, ASCE, Vol. 105, No. EM 4, Proc. Paper 14732, Aug. 1979, pp. 515-523.

346. Stathopoulos, T., "Wind Load on Eaves of Low Buildings," Journal of Structural Division, ASCE, Vol. 107, No. ST10, Proc. Paper 16543, Oct., 1981, pp. 1921-1934.

347. Columbus, J.K., "The Study of Pressure Coefficients on Large Flat Roofs and the Effect of Parapets on these Coefficients," B.E.Sc. Report, Faculty of Engineering Science, The University of Western Ontario, March, 1972.

348. Stathopoulos, T., "Wind Pressures on Low Buildings with Parapets", Journal of Structural Division, ASCE, Vol. 108, No. ST12, Proc. Paper 17578, Dec. 1982, pp. 2723-2736.

349. Wiren, B., "Wind Tunnel Investigation of the Pressure Distribution on a Flat Roof with Different Edge Profiles," translated from the Swedish, Library Communication 1606, Building Research Station, March, 1971.

350. Davenport, A.G., Surry, D. and Stathopoulos, T., "Wind Loads on Low-Rise Buildings: Final Report of Phase III - Parts 1 and 2," BLWT-SS4-1978, The University of Western Ontario, London, Canada, July 1978.

351. Peterka, J.A., and Cermak, J.E., "Adverse Wind Loading Induced by Adjacent Buildings, "Journal of Structural Division, ASCE, Vol. 102, March, 1976.

352. Zambrano, T.G., Peterka, J.A. and Cermak, J.E., "Wind-Load Interaction on an Adjacent Building," The Third U.S. National conference on Wind Engineering Research, The University of Florida, Gainesville, Febr. 1978.

353. Isyumov, N. and Davenport, A.G., "A Study of Wind Induced Exterior Pressures and Suctions on Lower Accommodation Levels of the CN Tower, Toronto," BLWT Report SS2-75, The University of Western Ontario, London, Canada.

354. Holmes, J.D. and Best, R.J., "A Wind-Tunnel Study of Wind Pressures on Grouped Tropical Houses," James Cook University of North Queensland, Wind Engineering Report 5/79, Queensland, Australia, Sept., 1979.

355. Lee, B.E., Hussain, M. and Soliman, B., "Predicting Natural Ventilation Forces Upon Low-Rise Buildings," ASHRAE Journal, Vol. 22, No. 2, Feb., 1980.

356. Baines, W.D., "Effects of Velocity Distribution on Wind Loads and Flow Patterns on Buildings," Proceedings of the Symposium on Wind Effects on Buildings and Structures, Vol. 1, National Physical Labs., Teddington, England, 1963.

357. Holmes, J.D., "Mean and fluctuating Internal Pressures Induced by Wind," James Cook University of North Queensland, Wind Engineering Report 5/78, Queensland, Australia, Dec., 1978.

358. Liu, H. and Saathoff, P.J., "Building Internal Pressure: Sudden Change," Journal of Engineering Mechanics, ASCE, Vol. 107, No. EM2, Proc. Paper 16172, April, 1981, pp. 309-321.

359. Saathoff, P.J. and Liu, H., "Predicting Internal Pressure of Multi-Room Buildings," Journal of Engineering Mechanics, ASCE, Vol. 109, No. 3, Paper 18028, June, 1983, pp. 908-919.

360. Minor, J.E. and Mehta, K.C., "Wind Damage Observations and Implications," J. of the Structural Division, ASCE, Vol. 105, No. ST11, Proc. Paper 14980, Nov. 1979, pp. 2279-2291.

361. Kandola, B.S., "Wind Effects on Buildings with Varying Leakage Characteristics - Wind Tunnel Investigation," Journal of Industrial Aerodynamics, Vol. 3, 1978, pp. 267-284.

362. Surry, D., Kitchen, R.B. and Davenport, A.G., "Wind Loading of two Hangars Proposed for Jeddah Airport, Saudi Arabia - A Rigid Model Study," Report to be published, Boundary Layer Wind Tunnel Laboratory, The University of Western Ontario, London, Canada.

363. Tamura, G.T., "Measurement of Air Leakage Characteristics of House Enclosures," ASHRAE Transaction, Vol. 81, Part I, 1975.

364. Newberry, C.W. and Eaton, K.J., "Wind Loading Handbook," Building Research Establishment Report, Department of the Environment, London, England, 1974.

365. Newberry, C. W., Eaton, K. J. and Mayne, J. R., "Wind Loading on Tall Buildings - Further Results from Royex House," Industrial Aerodynamics Abstracts 4, July - Aug., 1973.

366. Hazelwood, R.A., "Principles of Wind Loading on Tiled Roofs and their Application in the British Standard BS5534," Journal of Wind Engineering and Industrial Aerodynamics, Vol. 6, 1980, pp. 113-124.

367. Kind, R. J. and Wardlaw, R. L., "Wind Problems of Inverted Roofing Systems," Third Canadian Workshop on Wind Engineering, Canadian Wind Engineering Association, Vancouver, April, 1981, Toronto, May, 1981.

368. Stathopoulos, T. and Homma, H., "Natural Wind Effects on the Infiltration of Low-Rise Buildings," Proceedings of 8th CIB Congress, Oslo, Norway, June, 1980.

369. Liu, H. and Saathoff, P.J., "Internal Pressure and Building Safety," Journal of the Structural Division, ASCE, Vol. 108, No. ST10, Proc. Paper 17421, Oct., 1982, pp. 2223-2234.

370. Holmes, J.D. and Best, R.J., "Wind Tunnel Measurements of Mean Pressures on House Models and Comparison with Full Scale Data," James Cook University of North Queensland, Wind Engineering Report 3/77, Queensland, Australia, July, 1977.

371. Holmes, J. D. and Best, R. J., "Further Measurements on Wind Tunnel Models of the Aylesbury Experimental House," James Cook University of North Queensland, Wind Engineering Report 4/78,Queensland, Australia, October, 1978.

372. Apperley, L. Surry,. D., Stathopoulos, T. and Davenport, A.G., Comparative Measurements of Wind Pressure on a Model of the Full-Scale Experimental House at Aylesbury, England," Journal of Industrial Aerodynamics, Vol. 4, 1979, pp. 207-228.

373. Grennway, M.E. and Wood, C.J., "Wind Tunnel Pressure Measurements on the Aylesbury-Low-Rise Housing Estate," Part I, Simulation Design and Mean Pressures, University of Oxford, Dept. of Engineer. Science, Report No. 1213/77.

374. Greenway, M.E. and Wood, C.J., "Wind Tunnel Pressure Measurements on the Aylesbury Low-Rise Housing Estate," Part II, Mean RMS and Extreme Pressures with Frequency Spectra, University of Oxford, Dept. of Engineer. Science, Report No. 1271/78.

375. "American National Standard Building Code Requirements for Minimum Design Loads in Buildings and other Structures," ANSI Standard, American National Standards Institute A58.1-1982.

376. Stathopoulos, T., "PDF of Wind Pressures on Low-Rise Buildings," Journal of the Structural Division, ASCE, Vol. 106, No. ST5, Proc. Paper 15409, May, 1980, pp. 973-990.

377. "Aérodynamique," REEF-Volume II, Centre Scientifique et Technique du Bâtiment, april, 1980.

378. Wells, D.A. and Hoxey, R.P., "Measurements of Wind Loads on Full-Scale Glasshouses," Journal of Wind Engineering and Industrial Aerodynamics, Vols. 6, 1980, pp. 139-167.

379. "British Standard Code of Practice," CP3. Chapter V: Part 2: 1970, The Council for Codes of Practice British Standard Institution

380. Schweizerischer Ingenieur und Architekten Verein, Technische Normen No. 160, 1956.

381. "Wind Forces on Structures," Final Report of ASCE Task Committee on Wind Forces, Transactions, ASCE, Vol. 126, Part II, 1961, pp. 1124-1198.

382. Kareem, A. and Cermak, J.E., "Wind Forces and Moments on Prisma Flood Lighting Unit," Colorado State University, College of Engineering Report CER 75-76, AK-JEC27, 1976.

383. Tieleman, H.W., Akins, R.E. and Sparks, P.R., "An Investigation of Wind Loads on Solar Collectors," Virginia Polytechnic Institute and State University, VPI-E-80-1 Report, January 1980.

384. Simiu E. et al., "Wind, Earthquake, Snow, and Hail Loads on Solar Collectors," NBSIR 81-2199, National Bureau of Standards, U.S. Department of Commerce, Washington, DC, Jan. 1981.

385. Marchman III, J.F., "Wind Loading on Free Swinging Traffic Signals", Transportation Engineering Journal of ASCE, Vol. 97, No. TE2, Proc. Paper 8140, May, 1971, pp. 237-246.

386. Simiu, E., "Aerodynamic Coefficients and Risk Consistent Design", Journal of Structural Engineering, ASCE, Vol 109, No. 5, Proc. Paper 17951, May, 1983, pp. 1278-1289.

4. TALL BUILDINGS

Methods for designing tall buildings for wind loads have considerably evolved during the last few decades. This evolution has come about as a result of (1) the trend toward lighter structural systems that are more flexible and have less damping and can thus experience significant fluctuating motions and stresses, and (2) the increasing use in facades of cladding glass panels, for which wind loading is a primary design consideration.

The structural behavior of tall buildings subjected to wind loads depends on the extent to which the buildings are affected by the flow in the wake of other tall structures. It is therefore convenient to distinguish between buildings subjected to strong interference effects and buildings for which interference effects may be neglected in practice. As a very approximate guide, it may be assumed that if the distance between two tall buildings exceeds, say, eight times the average of the horizontal dimensions of the buildings, the mutual interference effects will be negligible for practical purposes. More refined guidelines can be obtained, for various specific situations, from studies concerning interference effects reported in references 401 through 405. A comparison was made in reference 401 between the response of square buildings located in urban terrain in the following two basic cases: (1) the building is located near a building with similar geometry and dimensions, and (2) there is no neighboring building causing significant interference effects. It was concluded that if the building is designed to withstand the loads (including the across-wind loads) it would experience in the second case, then it will perform satisfactorily in the first case as well, regardless of the relative position of the two buildings. In most cases, tall structures for which it is judged that interference effects may be significant are designed on the basis of specially conducted wind tunnel tests. However, in many instances interference effects are clearly minor, and it is possible to estimate the structural response by using procedures available in the literature.

Occupant comfort is widely believed to depend upon the acceleration experienced by the building under the action of fluctuating loads. Although this section reviews procedures for estimating building accelerations, no material is presented herein on serviceability (occupant comfort) criteria. A review of such criteria can be found in reference 111.

This chapter includes: a review of procedures for estimating along-wind, across-wind, and torsional response of structures for which interference effects may be neglected (sections 4.1, 4.2, and 4.3), a review of material on the use of tuned mass damper (TMD) devices for mitigating wind-induced building oscillations (section 4.4), and general information on design for wind-induced cladding loads (section 4.5).

4.1 ALONG-WIND RESPONSE

Following Liepmann's fundamental paper on the response of elastic structures immersed in turbulent flow, published in 1952 [407], procedures for estimating along-wind response were developed in reference 408 and subsequently in references 409 to 413. All these procedures are based on standard methods of structural dynamics as applied to the case of random excitations, and consist essentially of expressions derived by numerical integration from Eqs. 2.2.9 [111]. As shown in some detail reference 111, the procedure presented in reference 413 incorporates significant improvements over references 408, 409, and 410 in the modeling of the atmospheric flow and of the along-wind cross-correlation of the fluctuating pressures; has definite advantages over references 408 through 412 from a computational viewpoint; and yields results that are physically more realistic than the procedure incorporated in the Appendix to the American National Standard ANSI A58.1-1982 [119]. Like references 408 through 412, reference 413 is applicable to structures of the type represented in figure 4.1.1 for which (a) the fundamental mode shape may be approximated by a straight line, and (b) the contribution to the response of the vibration modes higher than the fundamental is negligible. Reference 413 also contains a procedure for estimating the along-wind response of "point structures". i.e., structures that may be viewed approximately as consisting of a single mass M concentrated at a height H (fig. 4.1.2).

For structures of the type represented in figure 4.1.1 but having a fundamental mode that differs significantly from a straight line, or having higher vibration modes that contribute significantly to the total deflection, the along-wind response may be estimated by using the computer program of reference 414.

For point structures (fig. 4.1.2) the procedure for estimating along-wind response consists of using eqs. 1a through 12a and 13 through 22 of table 4.1.1.

For buildings of the type represented in figure 4.1.1 the procedure consists of using eqs. 1b through 12b and 13 through 22 of table 4.1.1. The equivalent static wind load is equal to the mean load multiplied by the gust factor, G. In addition to the notations shown in figures 4.1.1 and 4.1.2 the following notations are used in table 4.1.1: z_o = roughness length (see table 1.2.1); z_d = zero plan displacement (z_d = 0 for all values of table 1.2.1), n_1 = natural frequency of vibration in fundamental mode of vibration, u_* = friction velocity (eq. 1.2.3), C_D = drag coefficient ($C_D = C_W + C_\ell$), C_W and C_ℓ = average pressure coefficients on windward and leeward face of building, respectively, M = total mass of structure with dimensions b, h, d (figure 4.1.2), z = height above ground, M(z) = mass of building per unit height (fig. 4.1.1), $\rho_b(z)$ = bulk mass of building per unit volume, ζ = damping ratio, ρ = mass of air per unit volume, ($\rho \simeq 1.25$ kg /m^3), β=coefficient given in table 1.2.5, T = duration of storm (T $\simeq$ 3,600 sec), $\overline{X}$ = mean displacement at top of structure, G = gust response factor, σ_x =

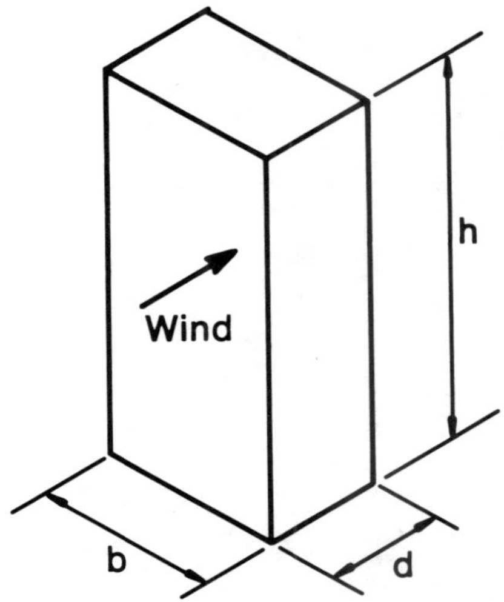

Figure 4.1.1 Notations

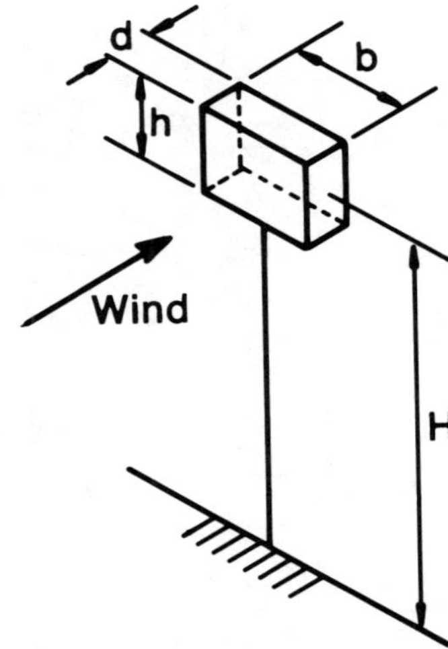

Figure 4.1.2 Notations

Table 4.1.1 Equations for Estimating the Along-Wind Response of Point Structures and of Buildings with a Linear Modal Shape (after Ref. 413).

(1a) $P = \ell n \frac{H - z_d}{z_o}$	(1b) $Q = 2(1 - \frac{z_d{}^2}{h^2})\ell n \frac{h - z_d}{z_o} - 1$
(2a) $J = 6.25\ P^2$	(2b) $J = 0.78\ Q^2$
(3a) $B = \frac{150\ P^2}{1.06 + 0.4 \frac{b}{H} + 0.25 \frac{h}{H} - 0.23 \frac{b}{H}\frac{h}{H}}$	(3b) $B = \frac{6.71\ Q^2}{1 + 0.26 \frac{b}{h}}$
(4a) $\tilde{f}_1 = \frac{n_1 H}{u_*}$	(4b) $\tilde{f}_1 = \frac{n_1 h}{u_*}$
(5a) $C(x) = \frac{1}{x} - \frac{1}{2x^2}(1 - e^{-2x})$	(5b) $C(x) = \frac{1}{x} - \frac{1}{2x^2}(1 - e^{-2x})$
(6a) $x_1 = 6.16 \frac{\tilde{f}_1}{P}\frac{d}{H}$	(6b) $x_1 = 12.32 \frac{\tilde{f}_1}{Q}\frac{d}{h}$
(7a) $N(\tilde{f}_1) = C(x_1)$	(7b) $N(\tilde{f}_1) = C(x_1)$
(8a) $C^2_{Df}(\tilde{f}_1) = C^2_w + 2C_w C_\ell N(\tilde{f}_1) + C^2_1$	(8b) $C^2_{Df}(\tilde{f}_1) = C^2_w + 2C_w C_\ell N(\tilde{f}_1) + C^2_1$
(9a) $x_2 = 2.56 \frac{\tilde{f}_1}{P}\frac{b}{H}$	(9b) $x_2 = 3.55 \frac{\tilde{f}_1}{Q}$
(10a) $x_3 = 1.60 \frac{\tilde{f}_1}{P}\frac{b}{H}$	(10b) $M(z) = bd\rho_b(z)$
(11a) $R = 9.40 \frac{P^2}{\zeta_1}(\frac{P}{\tilde{f}_1})^{2/3}\frac{C^2_{Df}(\tilde{f}_1)}{C^2_D} C(x_2)\ C(x_3)$	(11b) $R = 0.59 \frac{Q^2}{\zeta_1}(\frac{Q}{\tilde{f}_1})^{2/3}\frac{C^2_{Df}(\tilde{f}_1)}{C^2_D} C(x_2) \frac{1}{1 + 3.95 \frac{f_1}{Q}\frac{b}{h}}$
(12a) $M_1 = M$	(12b) $M_1 = \frac{1}{h^2}\int_o^h M(z)z^2 dz$

(13) $q_* = \frac{1}{2}\rho u_*{}^2$

(14) $\bar{X} = \frac{C_D b h q_*}{M_1(2\pi n_1)^2} J$

(15) $\sigma_X = \frac{C_D b h q_*}{M_1(2\pi n_1)^2}(\frac{\beta B}{6} + R)^{1/2}$

(16) $\nu_X = n_1(\frac{R}{\frac{\beta B}{6} + R})^{1/2}$

(17) $g_X = [1.175 + 2\ell n(\nu_X T)]^{1/2}$

(18) $G = 1 + g_X \frac{\sigma_X}{\bar{X}}$

(19) $X = G\bar{X}$

(20) $\sigma_{\ddot{X}} = \frac{C_D b h q_*}{M_1} R^{1/2}$

(21) $g_{\ddot{X}} = [1.175 + 2\ln(n_1 T)]^{1/2}$

(22) $\ddot{X} = g_{\ddot{X}}\ \sigma_{\ddot{X}}$

r.m.s. of fluctuating part of displacement at top of structure, $\sigma_{\ddot{x}}$ = r.m.s. acceleration at top of structure, $\ddot{X}$ = peak acceleration at top of structure. The estimation of the along-wind response of a tall building is illustrated below.

Numerical Example. Consider a building with $h = 200$ m; $b = 35$m; $d = 35$m; $n_1 = 0.175$ Hz; $\zeta = 0.01$; $\rho_b = 200$ kg/m^3; $C_W = 0.8$; $C_\ell = 0.5$; $C_D = 1.3$. The building is located in a town[a] (terrain category IV, see table 1.2.1, assumed to have homogeneous roughness over a distance upwind of at least 12.5h, see equation 1.2.9). It is assumed $\rho = 1.25$ kg/m^3, and that the 50-year fastest-mile wind speed at 10 m above ground in open (Category II) terrain is $u_{ofm}(10) = 78$ mph.

From eq. 1.2.10, the averaging time for the 50-year fastest mile wind speed is 3,600/78 = 46 sec. From figure 1.2.2, C(t) = 1.42, and from eqs. 1.2.17, 1.2.12, and 1.2.3, and tables 1.2.1 and 1.2.5, the hourly mean speed at 10 m above ground in open (category II) terrain is

$$u_{ref} = 78 - 1.42\ \beta^{1/2}\ \frac{u_{ref}}{2.5 \ln \frac{10}{0.07}}$$

or $u_{ref} = 62.3$ mph (27.8 m/s). From eq. 1.2.6 and table 1.2.4, the mean hourly wind speed at 10 m above ground in category IV terrain is $u(10) = 17.16$ m/s, and from eq. 1.2.3, $u_* = 2.98$ m/s. Then, referring to the equations of table 4.1.1, $Q = 9.60$ (eq. 1b); $J = 71.83$ (eq. 2b); $B = 591$ (eq. 3b); $\tilde{f}_1 = 11.74$ (eq. 4b); $x_1 = 2.62$ (eq. 6b); $N(\tilde{f}_1) = 0.31$ (eqs. 7b and 5b); $C^2_{Df}(\tilde{f}_1) = 1.14$ (eq. 8b); $x_2 = 4.34$ (eq. 9b); $M(z) \simeq 245{,}000$ kg (eq. 10b); $R = 353$ (eq. 11b); $M_1 = 16{,}333{,}300$ kg (eq. 12b); $q_* = 5.55$ kg/m/s2 (eq. 13); $\bar{X} = 0.184$ m (eq. 14); $\sigma_x = 0.074$ m (eq. 15); $\nu_x = 0.114$/sec (eq. 16); $g_x = 3.63$ (eq. 17); $G = 2.46$ (eq. 18); $X = 0.452$ (eq. 19); $\sigma_{\ddot{x}} = 0.058$ m/sec^2 (eq. 20); $g_{\ddot{x}} = 3.75$ (eq. 21); $\ddot{X} = 0.218$ m/sec^2 (eq. 22).

The procedures of references 409 through 414 estimate the response corresponding to the case where the mean wind is normal to a building face. Experiments reported in references 415 and 416 showed that the largest along wind deflections occur in this case.

4.2 ACROSS-WIND RESPONSE

The across-wind response of tall buildings is induced by vorticity shed in the wake of the building. As shown in reference 415, different expressions for the across-wind response of tall buildings are applicable depending upon whether or not the r.m.s. value of the across-wind oscillations at the tip of the building, σ_y,

---

[a] It is assumed that the terrain roughness is homogeneous over a distance upwind of at least 12.5 h (see eq. 1.2.9).

exceeds a critical value σ_{ycr}. If $\sigma_y > \sigma_{ycr}$, aeroelastic (lock-in) effects (see section 2.3.1) become significant and contribute to a considerable and dangerous increase of the across-wind response. Wind tunnel experiments reported in references 415 and 416 (p. 81) suggest that it is conservative from a structural design viewpoint to assume $\sigma_{ycr}/b \approx 0.015$, 0.025, and 0.045 in open, town, and city center exposure respectively (b = across-wind dimension of building).

For $\sigma_y < \sigma_{ycr}$ the across-wind response corresponding to the case where the mean wind is normal to a building face can be estimated by using the expressions [406]:

$$\sigma_y(h) \simeq \frac{\pi^{1/2}}{2\zeta^{1/2}} \frac{1}{(2\pi n_1)^2 M_1} \frac{1}{2} \rho u^2(h)\, bh\tilde{Y} \qquad (4.2.1)$$

$$\sigma_{\ddot{y}}(h) = (2\pi n_1)^2 \sigma_y \qquad (4.2.2)$$

where u(h) = mean wind speed at elevation h, $\tilde{Y}$ is a factor determined empirically, and the other notations are the same as in table 4.1.1. According to reference 417, the contribution to the response of modes higher than the fundamental is negligible in practice. For buildings with a uniformly distributed mass, a square shape in plan, and a fundamental modal shape that may be approximated by a straight line,

$$M_1 = \frac{1}{3} \rho_b\, b^2 h$$

and

$$\sigma_y(h) \simeq 0.0337 \left[\frac{u(h)}{n_1 b}\right]^2 \frac{\rho}{\rho_b} \frac{1}{\zeta^{1/2}} b\tilde{Y} \qquad (4.2.3)$$

Values of $\tilde{Y}$ based on measurements reported in references 415, 418, 419 and 420, and implicit in references 421 and 422, are given in table 4.2.1. Note that considerable differences exist among values of Y obtained from various sources (e.g., for urban terrain, nb/u(h) = 0.105 and b/h = 1/8.33, $\tilde{Y}$ = 10 according to references 419, and $\tilde{Y}$ = 23 on the basis of data from reference 418). Studies reported in reference 419 indicate that, for buildings with a square shape in plan, the across-wind response corresponding to cases where the wind is not normal to a building face is less than that given by eq. 4.2.1.

<u>Numerical Example.</u> The building previously considered is assumed to be acted upon by wind corresponding to a fastest-mile speed at 10 m above ground in open (category II) terrain $u_{ofm}(10)$ = 78 mph. As shown previously, the mean hourly wind speed at 10 m above ground in category IV terrain is u(10) = 17.16 m/s, or u(200) = 17.16[ln(200/1.00)]/[ln(10/1.00)] = 39.4 m/s. Then $n_1 b/u(h)$ = 0.155; b/h = 1/5.7; $\tilde{Y} \simeq 0.075$ (Table 4.2.1) σ_y = 0.23 m (eq. 4.2.3);

Table 4.2.1 Values $10^2\tilde{Y}$

		$n_1 b/u(h)$															
Terrain	b/h	0.04	0.05	0.06	0.07	0.08	0.09	0.10	0.105	0.115	0.125	0.14	0.16	0.18	0.20	0.25	0.30
Urban[a]	1/9	4.7	5.7	7.2	9.7	12.	17.	23.	24.	19.	13.	9.0	6.8	5.7	5.2	4.5	3.9
	1/8.33	5.3[b] 5.3[d]	6.3[b] 6.5[d]	7.8[b] 8.1[d]	11.[b] 11.[d]	13.[b] 13.[d]	17.[b] 14.[d]	23.[b] 12.[d]	23.[b] 10.[d]	19.[b] 7.7[d]	14.[b] 6.3[d]	10.[b] 5.3[d]	6.9[b] 4.7[d]	6.3[b] 3.9[d]	5.7[b] 3.6[d]	4.7[b] 3.2[d]	3.8[b] 2.9[d]
	1/6	8.4	9.7	11.	13.	15.	19.	21.	21.	20.	18.	15.	12.	10.	8.2	5.6	3.5
	1/4	6.2[b] 7.9[c]	7.6[b] 9.9[c]	9.3[b]	11.[b] 16.[c]	12.[b] 17.[c]	17.[b] 17.[c]	15.[b] 12.[c]	15.[b] 11.[c]	15.[b] 9.5[c]	13.[b] 8.8[c]	12.[b] 7.8[c]	9.6[b] 6.8[c]	13.[b] 5.9[c]	6.8[b] 5.7[c]	4.9[b] 4.9[c]	3.4[b] 4.0[c]
	1/3	3.9	5.5	7.1	8.2	9.1	14.	10.	10.	9.5	9.0	8.4	7.2	6.5	5.5	4.2	3.3
Suburban[a]	1/18	3.6	4.1	5.0	5.7	8.4	11.	22.	29.	22.	10.	6.0	3.6	2.8	2.4	2.0	1.7
	1/9	5.1	6.0	7.5	10.	13.	19.	28.	29.	13.	7.8	6.0	4.5	3.5	2.8	2.6	2.3
	1/6	5.3	6.9	8.9	12.	16.	21.	23.	23.	20.	14.	10.	7.7	6.1	4.5	2.8	1.7
	1/4	5.0[b]	6.2[b]	8.0[b]	11.[b]	15.[b]	18.[b]	19.[b]	18.[b]	16.[b]	12.[b]	9.5[b]	7.4[b]	5.9[b]	4.7[b]	3.7[b]	2.1[b]
	1/3	4.6	5.4	7.1	10.	13.	15	15.	13.	12.	10.	8.9	7.1	5.7	4.9	3.5	2.6
	1/4	6.2[c]	7.1[c]	8.7[c]	13.[c]	17.[c]	28.[c]	29.[c]	28.	11.[c]	8.7[c]	7.5[c]	6.6[c]	5.9[c]	5.5[c]	4.8[c]	4.0[c]
All Terrains[e]	1/3.4–1/7.1	56.	39.	30.	24.	20.	16.	14.	13.	11.	10.	8.5	7.0	5.8	5.0	3.6	2.7
All Terrains[f]	1/3.4–1/7.1	29.	22.	17.	14.	12.	10.	8.9	8.3	7.1	6.6	5.9	4.9	4.1	3.6	2.8	2.1

[a] Results reported in Refs. 415 and 418, unless otherwise noted.

[b] Values obtained by interpolation between results reported in Ref. 418 for b/h = 1/6 and b/h = 1/3.

[c] Results reported in Ref. 420.

[d] Results reported in Ref. 419.

[e] Values implicit in Ref. 421. Experimental values may differ from tabulated values by ± 40%. Application restricted to buildings with density of the order of 200 kg/m^3 and damping ratio of the order of 1%. According to Ref. 421, these values may be applied to building with cross-sections other than square, i.e., triangular, rectangular and circular.

[f] Values implicit in Ref. 422.

$\sigma_{\ddot{y}} = 0.28 m/sec^2$ (eq. 4.2.2). Assuming that the peak factors are $g_y = 3.5$, $g_{\ddot{y}} = 4.0$, it follows that the peak across-wind response and acceleration are Y = 0.805 m and $\ddot{Y} \simeq 1.12 m/sec^2$. These values are larger than the corresponding values of the along-wind response calculated previously, i.e., $X \simeq 0.452$ m and $\ddot{X} \simeq 0.218 m/sec^2$.

The peak accelerations calculated in this example are relatively high, i.e., they would be rated as very annoying or intolerable from an occupant comfort point of view [111]. Note, however, that these accelerations correspond to the 50-year wind, and that it is common to judge the acceptability of building accelerations on the basis of shorter return periods, e.g., 10 years or less [111].

4.3 TORSIONAL RESPONSE

Wind-induced torsional effects occur because the center of mass and/or the elastic center do not coincide with the instantaneous point of application of the aerodynamic loads. Early efforts to develop design criteria for wind-induced torsional effects were reported in reference 423 and 424. A first attempt to study analytically torsion induced on buildings by fluctuating wind loads was reported in reference 425.

More recently, potentially useful methods for estimating the along-wind, across-wind, and torsional response of rectangular buildings were presented in references 426 and 427. However, owing to the absence of sufficient information on aerodynamic loads, the methods are not presently usable for design purpose.

Wind tunnel and full-scale research studies of torsional response were first reported in [401] and [428]. Reference 401 includes information on wind-induced torsional moments in an isolated square building model having a height to width ratio h/b = 8.33 in flow that simulates urban conditions. According to the results of reference 401, torsional moments are largest when the mean wind velocity is normal to a building face. As the angle, α, between the mean wind velocity and the normal to the building face increases from 0° to 45° the torsional moments decrease from their maximum value corresponding to $\alpha = 0$ to about 25 percent of that value for $\alpha = 45^o$. Assuming that the mechanical properties of the model are similar to those of typical high rise structures, it was estimated in [401] that, for $\alpha = 0$, the peak torsion-induced response of a corner column is approximately 65 percent of the peak along-wind response corresponding to $\alpha = 0$. For $\alpha = 45^o$, the peak torsion-induced response of a corner column is about 15 percent of the peak along-wind response corresponding to $\alpha = 0$.

Systematic wind tunnel studies conducted at the University of Western Ontario were subsequently reported in [429] to [423]. These studies have led to the following empirical relation for estimating the peak base torque $T_{max}[u(h)]$ induced by winds with speed u(h) at the top of the building:

$$T_{max}[u(h)] = \psi \left\{\overline{T}[u(h)] + g_T T_{rms}[u(h)]\right\} \quad (4.3.1)$$

where ψ is a reduction coefficient that is briefly discussed subsequently, $g_T \simeq 3.8$ is a torsional peak factor, and the linear and r.m.s. base torque, $\overline{T}[u(h)]$ and T_{rms} [u(h)] are given by the expressions

$$\overline{T}\,[u(h)] \simeq 0.038\rho L^4\, h\, n_T^2\, U_r^{\,2} \tag{4.3.2}$$

$$T_{rms}\,[u(h)] \simeq 0.00167\frac{1}{\zeta_T^{1/2}}\,\rho L^4\, h\, n_T^2\, U_r^{\,2.68} \tag{4.3.3}$$

$$U_r = \frac{U(h)}{n_T L} \tag{4.3.4}$$

$$L = \frac{\int |r|\, ds}{A^{1/2}} \tag{4.3.5}$$

In eqs. 4.3.2 to 4.3.5, ρ is the air density ($\rho \simeq 1.25$ kg/m^3), h is the height of the building, n_T and ζ_T are the natural frequency and the damping ratio in the fundamental torsional mode of vibration, ds is the elemental length of the building perimeter, $|r|$ is the torque arm of the element ds (i.e., the distance between the elastic center and the normal to the building boundary at the center of ds), and A is the cross-sectional area of the building. Equations 4.3.2 and 4.3.3 are based upon the experimental results shown in figures 4.3.1 and 4.3.2, in which the ordinates are the reduced mean and r.m.s base torque, $T_r = \overline{T}/(\rho L^4\, h\, n_T)$ and

$\sigma_r = T_{rms}\, \zeta_T^{1/2}/(\rho L^4\, h\, n_T^2)$, respectively. The torques $\overline{T}$ and T_{rms} are each induced by wind with reduced speed U_r and with the respective most unfavorable direction. In general, the most unfavorable directions for $\overline{T}$ and T_{rms} do not coincide. In addition, in most cases neither of these directions will coincide with the direction of the extreme winds expected to occur at the site. For these reasons, the coefficient ψ in eq. 4.3.1 is less than unity. It is estimated in [429] that $0.75 < \psi \leq 1$ in most cases.

The peak torsional-induced horizontal accelerations at the top of the building at the distance v from the elastic center can be written as:

$$\ddot{\theta}\, v \simeq 2g_T\, T_{rms}\, v/(\rho_b\, bd\, h\, r_m^{\,2}) \tag{4.3.6}$$

where $\ddot{\theta}$ is the peak angular acceleration and r_m is the radius of gyration. For a rectangular shape with uniform bulk mass per unit volume ρ_b,

$$r_m = \frac{b^2 + d^2}{12} \tag{4.3.7}$$

Eq. 4.3.6 was obtained in [429] assuming a linear fundamental modal shape, a uniform building bulk mass per unit volume, and negligible contributions by higher torsional modes of vibration.

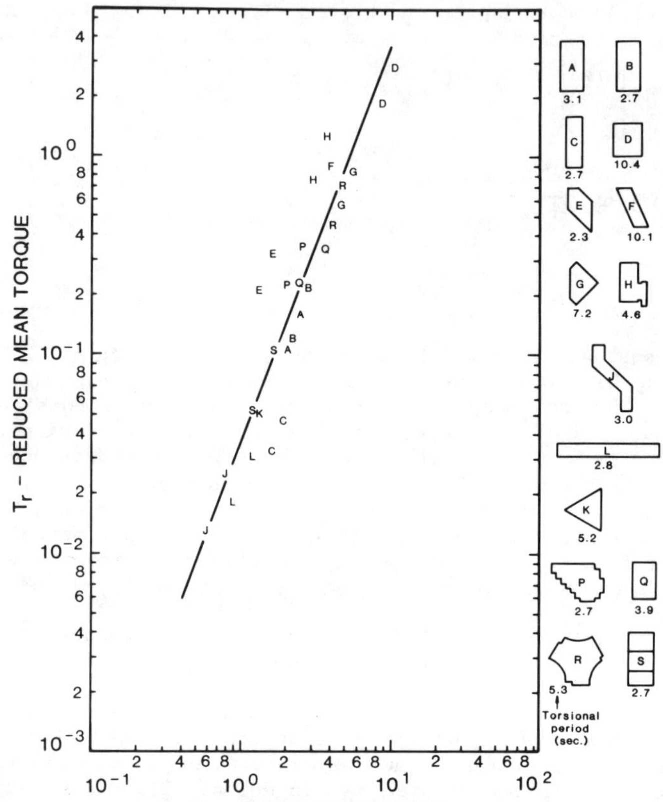

Figure 4.3.1 - Reduced Mean Torque as Measured for Buildings with Various Shapes in Plan.

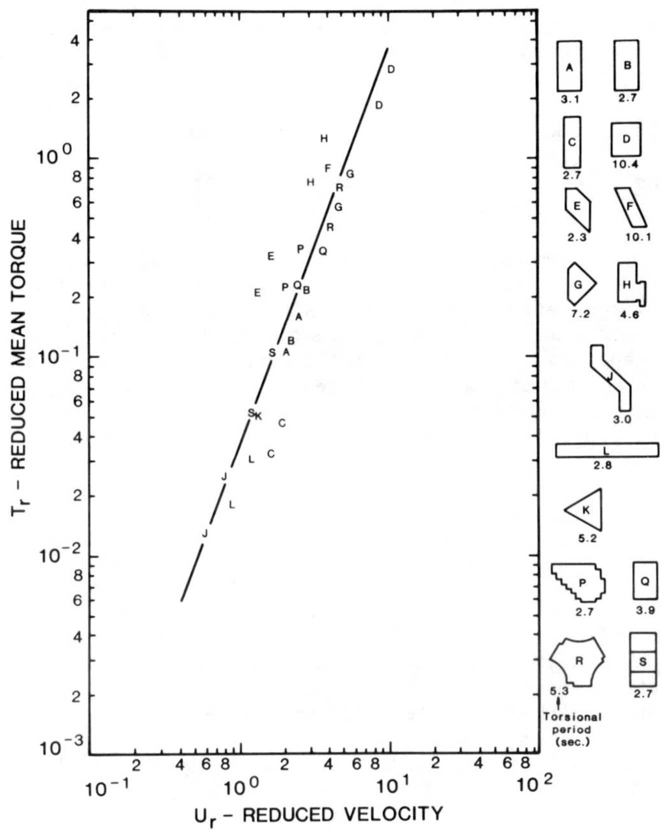

Figure 4.3.2 Reduced r.m.s. Torque as Measured for Buildings with Various Shapes in Plan.

Numerical Example. For the building considered previously h = 200m, b = d = 35 m, u(h) = 39.4 m/s, ρ_b = 200 kg/m^3. It is assumed that the natural frequency and the damping ratio in the fundamental torsional mode of vibration are n_T = 0.3 Hz and ζ_T = 0.01, respectively, and that the air density is ρ = 1.25 kg/m^3.

From eq. 4.3.5, L = 8(b/2)(b/4)/b = 35 m. Then, U_r = 39.4/(0.3 x 35) = 3.75 (eq. 4.3.4), $\overline{T}$[39.4] = 1.8 x 10^7 Nm (eq. 4.3.2), T_{rms} [39.4] = 1.95 x 10^7 Nm (eq. 4.3.3), T_{max} = 9.2 x 10^7 Nm (eq. 4.3.1) in which it is assumed ψ = 1, $g_T \simeq$ 3.8). The peak torsion-induced horizontal acceleration at the top corner (r = 35 x $\sqrt{2}/2$ = 24.7 m) is $\ddot{\theta}_{max} r$ = 0.27 m/s^2. Note that this exceeds the peak along-wind accelerations calculated in the previous numerical example.

The peak combined effect of the along-wind, across-wind, and torsional loads can be obtained by summing up vectorially the individual peak effects of these loads and multiplying the result by a reduction factor (equal to 0.8, say) which accounts for the fact that, in general, individual peaks do not occur simultaneously. If the combined effect so calculated is less than an individual effect, it is the latter that should be considered in design. A useful study of the effects of eccentricities of the center of mass and the elastic center is reported in reference 439.

4.4 TUNED MASS DAMPERS

A tuned mass damper (TMD) is a system consisting essentially of a mass, a spring and an energy dissipation device (dashpot). Its function is to reduce the amplitude of response oscillations of a body to which it is attached (fig. 4.4.1). The effect of the TMD is to create a resonant amplification curve (fig. 2.2.2) with a spike that is broader and flatter for the joint body TMD-system than for the original body not provided with a TMD.

In computational terms, the effect of the TMD may be viewed as changing the damping of the structure from the value ζ_1, that would be applicable in the absence of the TMD, to an increased value ζ_e. To the extent that the reliability of the TMDs is uncertain, TMD effects are counted upon for serviceability (occupant comfort) purposes only, and are disregarded in structural design [433].

Let the generalized mass of a building, the natural frequency, and the damping ratio in the fundamental mode of the building be denoted by M_1, n_1, and ζ_1, respectively, and let the mass, natural frequency, and damping ratio of the TMD system be denoted by M_2, n_2, and ζ_2, respectively.

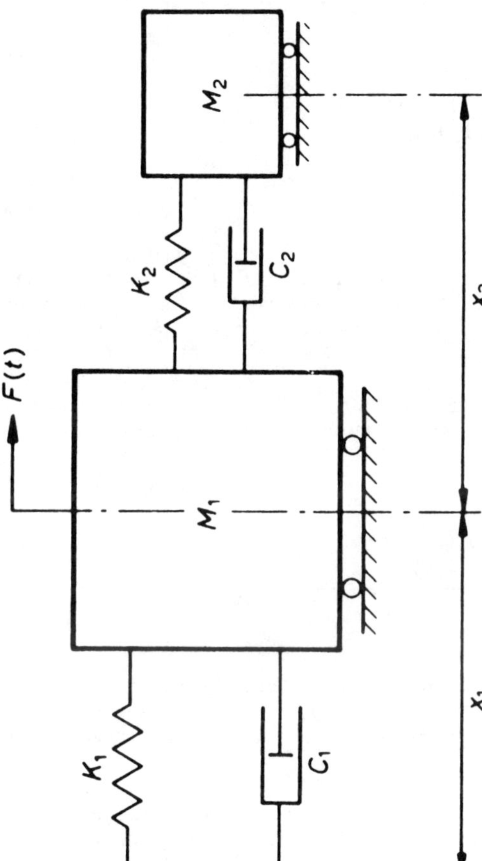

Figure 4.4.1 - Notations

Using results from references 434 and 435 it can be shown that

$$\zeta_e = \frac{1}{2}\,\frac{\alpha_1(\alpha_2\alpha_3 - \alpha_1) - \alpha_o\alpha_3^2}{\alpha_o(\alpha_2\alpha_3 - \alpha_1) + \alpha_3(\beta_1^2 - 2\alpha_o) + \alpha_1}$$

$$\begin{aligned}
\text{where } \alpha_o &= f^2 && (4.4.1a)\\
\alpha_1 &= 2f(\zeta_1 f + \zeta_2) && (4.4.1b)\\
\alpha_2 &= 1 + f^2(1+\mu) + 4f\zeta_1\zeta_2 && (4.4.1c)\\
\alpha_3 &= 2\zeta_1 + 2\zeta_2 f(1+\mu) && (4.4.1d)\\
\beta_1 &= 2\zeta_2 f && (4.4.1e)\\
\mu &= M_2/M_1 && (4.4.1f)\\
f &= n_2/n_1 && (4.4.1g)
\end{aligned}$$

For example, if $\mu = 0.01$, $f = 0.98$, $\zeta_1 = 0.01$, and $\zeta_2 = 0.0515$, $\zeta_e = 0.03226$. From eq. 4.4.1 it is possible to obtain the optimum (largest) value of ζ_e, ζ_e^{opt}, consistent with any given set of values ζ_1, f, μ. To ζ_e^{opt} there corresponds a value of ζ_2, denoted by ζ_2^{opt}.

Let the displacement of the TMD mass with respect to mass M_1 be denoted by x_2. It can be shown that the mean square value of x_2 can be expressed as:

$$\overline{x_2^2} = \frac{2\zeta_1\alpha_1}{\alpha_1(\alpha_2\alpha_3 - \alpha_1) - \alpha_o\alpha_3^2}\,\overline{x_{1,0}^2} \qquad (4.4.2)$$

where $\overline{x_{1,0}^2}$ = variance of wind-induced displacement of mass M_1 due to resonant amplification effects only. It is shown in reference 435 that the order of magnitude of x_2 may be as high as 1m in practical situations.

Eqs. 4.4.1 and 4.4.2 do not take into account the multi-degree of freedom characteristics of tall buildings, and should hence be regarded as approximate. The errors involved in modeling a building by the single mass M_1 are in practice small (of the order of 10 percent or less). Reference 435 contains material directly applicable to the study of these errors.

In some situations, TMD's can be designed to reduce torsion-induced motions. The procedures for carrying out the pertinent calculations are similar to those described in this section. Additional information on TMDs and their application to tall buildings can be found in references 437 through 443.

4.5 CLADDING LOADS

As pointed out in Chapter 2, peak wind loads experienced by building parts or portions during any given storm depend upon the size of the area upon which they act, i.e., they tend to be larger for smaller areas, and vice-versa. Provisions concerning the magnitude of peak wind loads experienced by parts and portions, including cladding panels in tall buildings, are set forth in the American National Standard A58.1 [119] as a function of position on the structure. These provisions cannot cover the great variety of aerodynamic situations that may arise in practice, particularly as far as the relatively high local loads are concerned that occur near zones of flow separation (see section 2.1), i.e., near building corners and eaves. For this reason, it has become increasingly common practice to design cladding on the basis of test results obtained in wind tunnels that attempt to simulate the atmospheric flow. From such results and standard data concerning internal pressures (e.g., ref. 119) cladding loads (averaged over 1 to 3 seconds) are estimated. These are transformed into 1-min loads that are nominally equivalent to the peak loads from the point of view of the effect on the glass strength. The 1-min loads are then used in conjunction with load tables or diagrams for the selection of cladding thickness as a function of material and of panel dimensions (e.g., ref. 444).

Laboratory tests indicate that, in some instances, peak suctions--which depend strongly upon the aerodynamics of the building-- are highest near the ground. While information based on such tests has been incorporated in reference 119, this trend is not reflected in some other codes and standards. Designs based on provisions of such codes and standards may therefore be unconservative in certain instances.

Wind tunnel tests show that wind pressures on cladding may depend very strongly upon the direction of the mean wind. For this reason, tall building models are subjected to tests wherein pressure coefficients are obtained for as many as 24 or more different directions. If the orientation of the building is known, and if information on the directions distribution of the extreme wind speeds is available, it is possible to integrate the direction-dependent aerodynamic and climatological features of the loading (see Section 1.3.1.3). Such integration has the effect of reducing, in many cases substantially, the estimated cladding loads corresponding to any specified mean recurrence interval. On the other hand, procedures for estimating cladding loads that ignore wind directionality effects result in nonuniform safety levels with respect to cladding failure over the surface of the building facade. Practical procedures for taking wind directionality into account in cladding design are given in references 445 through 449. Reference 111 describes a risk-consistent procedure for analyzing glass facades to meet specified safety performance criteria.

Useful information on the probability distributions of cladding pressure and suction peaks is presented in reference 450. Selected data on increased cladding pressures due to the proximity of adjacent buildings are reported in reference 405.

REFERENCES

401. Reinhold, T. A., et al., Mean and Fluctuating Forces and Torques on a Tall Building Model of Square Cross-Section as a Single Model, in the Wake of a Similar Model, and in the Wake of a Rectangular Model, Report VP1-E-79-11, Department of Engineering Science and Mechanics, Virginia Polytechnic Institute, Blacksburg, VA, March 1979.

402. Saunders, J. W., and Melbourne, W. H., "Buffeting Effects of Upstream Buildings," Proceedings, Fifth International Conference on Wind Engineering, Ft. Collins, Colorado, July 1979, Vol. I, Pergamon Press, 1980.

403. Blessman, J. and Pierce, J. D., "Interaction Effects in Neighboring Tall Buildings," Proceedings, Fifth International Conference on Wind Engineering, Ft. Collins, Colorado, July 1979, Vol. II, Pergamon Press, 1980.

404. Ruscheweyh, H., "Dynamic Response of High-Rise Buildings under Wind Action with Interference Effects from Surrounding Builidngs of Similar Size," Proceedings, Fifth International Conference on Wind Engineering, Ft. Collins, Colorado, July 1979, Vol. II, Pergamon Press, 1980.

405. Peterka, J. A. and Cermak, J. E., "Adverse Wind Loading Induced by Adjacent Buildings, Journal of the Structural Division, ASCE, Vol. 102, No. ST3, March, 1976, pp. 533-548.

406. Simiu, E., "Modern Developments in Wind Engineering - Part IV," Engineering Structures, Vol. 5, Oct. 1983, pp. 273-281.

407. Liepmann, H. W., "On the Application of Statistical Concepts to the Buffeting Problem," Journal of Aeronautical Sciences, Vol. 19, No. 12, December 1952, pp. 793-800 and 822.

408. Davenport, A. G., "The Application of Statistical Concepts to the Wind Loading of Structures," Proceedings, Institution of Civil Engineers, Vol. 19, 1961, pp. 449-472.

409. Davenport, A. G., "Gust Loading Factors," Journal of the Structural Division, ASCE, Vol. 43, No. ST3, June 1967, pp. 11-34.

410. Vellozzi, J. and Cohen, E., "Gust Response Factors," Journal of the Structural Division, ASCE, Vol. 94, No. ST6, June 1968, pp. 1295-1313.

411. Soize, C., "Dynamique stochastique des structures elancees soumises aux charges du vent," Revue Francaise de Mecanique, No. 60, 1976.

412. Simiu, E., "Revised Procedure for Estimating Along-wind Response, Journal of the Structural Division, ASCE, Vol. 106, No. ST1, January 1980, pp. 1-10.

413. Solari, G., "Along-wind Response Estimation: Closed Form Solution," Journal of the Structural Division, ASCE, Vol. 1108, No. ST1, January 1982, pp. 225-224.

414. Simiu, E. and Lozier, D. W., "The Buffeting of Structures by Strong Winds - Windload Program," NTIS Accession No. PB 294757/AS, Computer Program for Estimating Along-Wind Response, National Technical Information Service, Springfield, VA, 1979.

415. Kwok, C. S. and Melbourne, W. H., "Wind-Induced Lock-In Excitation of Tall Structures," Journal of the Structural Division, ASCE,Vol. 107, No. St1, January 1981, pp. 57-72.

416. Rosati, P. A., An Experimental Study of the Response of a Square Prism to Wind Load, BLWT II-68, Faculty of Graduate Studies, University of Western Ontario, London, Ontario, Canada, 1968.

417. Kareem, A., "Wind-Excited Response of Buildings in Higher Modes," Journal of the Structural Division, ASCE, Vol. 107, No. ST, April 1981, pp. 701-706.

418. Saunders, J. W., Wind Excitation of Tall Buildings with Particular Reference to the Cross-Wind Motion of Tall Buildings of Constant Rectangular Cross-Section. Thesis presented to the Department of Mechanical Engineering, Monash University, at Victoria, Australia in 1975, in fulfillment of the requirements for the degree of Doctor of Philosphy.

419. Reinhold, T. A. and Sparks, P. R., "The Influence of Wind Direction on the Response of a Square-Section Tall Building," Proceedings. Fifth International Conference on Wind Engineering, Fort Collins, Colorado, July 1979, Pergamon Press, 1980.

420. Kareem, A., "Cross-Wind Response of Buildings," Journal of the Structural Division, ASCE,Vo. 108, April 1982, pp. 869-887.

421. Vickery, B. J., "Notes on Wind Forces on Tall Buildings," Annex to Australian Standard 1170, Part 2-1973, SAA Loading Code Part 2 - Wind Forces, Standards Association of Australia, Sydney, 1973.

422. Commentaries on Part 4 of the National Building Code of Canada 1975, Supplement No. 4 to the National Building Code of Canada, NRCC No. 13989, Associate Committee of the National Building Code, National Research Council of Canada, Ottawa, 1975.

423. Smith, A., "Basis of Design for Hurrican Exposure", Report of Committee 308, Proceedings, ACI, Vo. 27, 1931, p. 903.

424. "Torsional Effects of Wind in Buildings," in Wind Bracing in Steel Buildings, Sixth Progress Report of Subcommittee No. 31, Committee on Steel of the Structural Division, Reports, ASCE, June 1939, pp. 988-996.

425. Patrickson, C. P. and Friedmann, P., A Study of the Coupled Lateral and Torsional Response of Tall Buildings to Wind Loadings, School of Engineering and Applied Science, University of California, Los Angeles, UCLA-ENG-76126, December 1976.

426. Foutch, D. A. and Safak, E., "Torsional Vibration of Wind-Excited Symmetrical Structures", Journal of Wind Engineering and Industrial Aerodynamics, 7 (1981) 191-201.

427. Foutch, D.A. and Safak, E., "Torsional Vibration of Along-Wind Excited Structures", Journal of the Engineering Mechanics Division, ASCE, 107, (1981) 323-337.

428. Hart, G. C., DiJulio, R. M., and Lew, M., "Torsional Response of High-Rise Buildings", Journal of the Structural Division, ASCE, 101 (1975) 323-337.

429. Greig, G. L., Toward an Estimate of Wind-Induced Dynamic Torqueon Tall Buildings, Master's thesis, Dept. of Engineering. University of Western Ontario, London, Ont., Sept. 1980.

430. Surry, D. and Lythe, G., "Mean Torsional Loads on Tall Buildings", Proceedings, Fourth U. S. National Conference on Wind Engineering Research, Dept. of Civil Eng., University of Washington, Seattle, Washington, July 1981.

431. Isyumov, N., "The Aeroelastic Modeling of Tall Buildings", Proceedings, International Workshop on Wind Tunnel Modeling for Civil Engineering Applications, Cambridge Univ. Press., 1982.

432. Isyumov, N. and Poole, M., "Wind-Induced Torque on Square and Rectangular Building Shapes", Proceedings, Sixth International Conference on Wind Engineering, Elsevier Scientific Publishing Company, Amsterdam, 1984.

433. Wiesner, K.B., "Tuned Mass Dampers to Reduce Building Wind Motion," Preprint 3510, ASCE Convention and Exposition, Boston, April 2-6, 1979.

434. Crandall, S. H. and Mark, W. D., Random Vibrations in Mechanical Systems, Academic Press, Inc., New York, NY, 1963.

435. McNamara, R. J., "Tuned Mass Dampers for Buildings," Journal of the Structural Division, ASCE, Vol. 103, No. ST9, Proceedings Paper No. 13200, September 1977, pp. 1785-1798.

436. Der Kiureghian, A., Sackman, J. L., and Nour-Omid, B., Dynamic Response of Light Equipment in Structures, Report UCB/EERC-81/05, Earthquake Engineering Research Center, College of Engineering, University of California, Berkeley, California, April 1981.

437. Der Hartog, J. P., Mechanical Vibrations, McGraww-Hill Book Company, Inc., New York, NY 1956.

438. Isyumov, N., Holmes, J. and Davenport, A. G., A Study of Wind Effects for the First National City Corporation Project - New York, U.S.A., University of Western Ontario Research Report BLWT-551-75, London, Ontario, Canada, 1975.

439. Luft, R. W., "Optimal Tuned Mass Dampers for Buildings," Journal of the Structural Division, ASCE, Proceedings Paper 15017, December 1979, pp. 2766-2772.

440. "Tower Cables Handle Wind, Water Tank Damps It," Engineering News Record, December 9, 1971, p. 23.

441. "Lead Hula-Hoops Stabilize Antenna," Engineering News Record, Vol. 197, No. 4, July 22, 1976, p. 23.

442. "Hancock Tower Now to Get Dampers," Engineering News Record, October 30, 1975, p. 11.

443. "Tuned Mass Dampers Sway Skyscrapers in Wind," Engineering News Record, August 18, 1977, pp. 28-29.

444. Architectural Glass Products, PPG Industries, Pittsburgh, PA, Jan. 1979.

445. Tryggvason, B. J., Surry, D., and Davenport A. G., "Predicted Wind-Induced Response in Hurricane Zones," Journal of the Structural Division, ASCE, Vol. 102, No. ST12, Proceedings Paper 12630, December 1976, pp. 2333-2350.

446. Mayne, J. R. and Cook, N. J., On Design Procedures for Wind Loading, Building Research Establishment Current Paper CP 25/78, Department of the Environment, Great Britain, 1978.

447. Simiu, E. and Filliben, J. J., "Wind Direction Effects on Cladding and Structural Loads," Engineering Structures, July 1981, pp. 181-186.

448. Simiu, E. and Batts, M. E., "Wind-Induced Cladding Loads in Hurricane-prone Regions", Journal of Structural Engineering, ASCE, Vol. 109, Jan. 1983, pp. 262-266.

449. Simiu, E., "Aerodynamic Coefficients and Risk-Consistent Design," Journal of Structural Engineering, Vol. 109, May 1983, pp. 1278-1289.

450. Peterka, J. A. and Cermak, J. E., "Wind Pressure on Buildings-Probability Densities," Journal of the Structural Division, ASCE,Vol. 101, June 1975, pp. 1255-1267.

451. Tallin, A., and Ellingwood, B., "Wind Induced Lateral-Torsional Motion of Buildings", Journal of Structural Engineering, Vol. III, No. 10, October 1985, pp. 2197 - 2213.

5. SLENDER TOWERS AND STACKS

Slender towers and stacks must be designed to withstand the effects of both along-wind and across-wind loads.

The along-wind response can be estimated by using the procedures referred to in section 4.1 in which appropriate values for the pressure coefficients C_ℓ and C_w are used. In the case of structures with a circular shape in plan, it may be assumed for design purposes $C_w \simeq 1.2$, $C_\ell \simeq 0$ ($Re < 3 \times 10^5$), and $C_w \simeq 0.8$, $C_\ell = 0$ ($Re > 3 \times 10^5$). In reality C_w depends also on the structure's aspect ratio and surface roughness. Information on the effect of these two factors is compiled in reference 501 and summarized in reference 111.

5.1 Across-wind Response

The character of the across-wind response depends on whether the across-wind deflections are sufficiently large for aeroelastic (lock-in) effects to occur (see section 2.3.1). Basic studies of wind-induced oscillations involving lock-in effects are reported in references 501 through 517 and are reviewed in references 111, 518 and 519.

Owing to Reynolds number effects, extrapolation of wind tunnel test results to the prototype can be misleading, unless complemented by careful studies based on full scale aerodynamic and aeroelastic data [529].

Note that structures should be designed for the simultaneous occurrence of the along-wind and across-wind response. Individually the peak fluctuating part of the along-wind response and the peak across-wind response for structures that do not experience lock-in effects may be assumed to be 3.5 to 4.0 times the respective r.m.s. response. However, because the probability of simultaneous occurrence of these peaks is low, it is reasonable to assume that, unless the response in either direction is dominant, the peak of the total response is only about 2.8 to 3.2 times the total r.m.s. response.

5.1.1 Rumman's Procedure [520]

In this procedure it is assumed that towers or stacks with a circular cross-section are subjected to a sinusoidal force per unit length

$$F_o(z) = \frac{1}{2} C_L \rho u^2(z_{ei}) D(z) \tag{5.1.1}$$

where ρ = density, $u(z_{ei})$ = mean wind speed at elevation z_{ei}, C_L = lift coefficient, and $D(z)$ = diameter of structure at elevation z. Reference 520 suggests $z_{ei} = h$, where h = height of structure. Other practitioners suggest z_{ei} = 2/3 h to 5/6 h [521], or z_{ei} = 2/3 h [522]. The wind speeds being considered are those producing at elevation z_{ei} vortex shedding with frequencies equal to the natural frequencies of the structure, i.e.,

$$u(z_{e_i}) = \frac{1}{S} n_i D(z_{e_i}) \qquad (i = 1,2 \ldots) \tag{5.1.2}$$

(see eq. 2.1.7). For Reynolds numbers based on $u(z_{ei})$ and $D(z_{ei})$ such that $Re > 2 \times 10^6$ or so it is usually assume $S \simeq 0.20$ to 0.25. From eqs. 2.2.2, 2.2.3, 5.1.1 and 5.1.2 it follows that the peak response for the structure excited in the i-th mode may be written as

$$y_{pk}(z) = \frac{\rho}{16\pi^2} \frac{C_L}{\zeta_i} D^2 (z_{e_i}) \frac{\int_o^h D(z) y_i(z) dz}{M S_i^2} y_i(z) \tag{5.1.3}$$

where $y_i(z)$ = i-th normal mode of vibration, ζ_i = damping in i-th mode, and M_i is the generalized mass (eq. 2.2.6). According to reference 520 many tall reinforced concrete chimneys were designed using a ratio $C_L/\zeta_1 \simeq 13$ to 16.

According to reference 523 it was determined from observations that tall reinforced concrete chimneys with constant or nearly constant diameter do not appear to experience unacceptably large motions if their Scruton number c_i, defined as

$$c_i = \frac{2M_i}{\int_o y_i^2(z) dz} \quad \frac{\zeta_i}{\rho D^2(z_{ei})} \tag{5.1.4}$$

is larger than four (see also reference 523).

5.1.2 Procedures by Vickery and Basu [526 to 528]

It is generally agreed that the loading and response models inherent in Rumman's procedure are not entirely consistent with advances made over the last two decades in the fields of micrometerology, aerodynamics and structural dynamics. This could lead, in certain situations, to the underestimation of across-wind response [524].

In reference 524, Vickery and Clark attempted the development of a conceptually more advanced procedure, a corrected version of which is presented in detail in reference 525. This procedure takes into account the variation of wind speed with height in the atmospheric boundary layer and makes use of spectral (frequency domain) techniques in the estimation of the across-wind response. Further research led to the development of improved procedures in references 526 to 528.

The r.m.s. across-wind response in the i-th mode of vibration may for practical purposes be estimated as follows:

$$\sigma_{yi}(z) = \overline{\xi_i^2}^{\,1/2}\; y_i(z) \qquad (5.1.5)$$

$$\overline{\xi_i^2} \simeq \overline{\xi_{oi}^2}\; \frac{\zeta_i}{\zeta_i + \zeta_{ai}} \qquad (5.1.6)$$

where ζ_i = structural damping, ζ_{ai} = aerodynamic damping, and $y_i(z)$ = i-th modal shape.

For <u>structures with uniform circular cross-section</u>

$$\overline{\xi_{oi}^2}^{\,1/2} \simeq \frac{0.035\; \overline{C_L^2}^{\,1/2}\; L^{1/2}}{\zeta_i^{1/2}\; S^2}\; \frac{\rho D^3}{M_i}\; \left[D \int_o^h y_i^2(z)dz\right]^{1/2} \qquad (5.1.7)$$

$$\zeta_{ai} = -\frac{\rho D^2}{M_i}\; K_{ao} \int y_i^2(z)dz \qquad (5.1.8)$$

[528] where ρ = air density, M_i = generalized mass in the i-th mode, D = outside diameter $\overline{C_L^2}^{\,1/2}$ = r.m.s. lift coefficient, L = spanwise lift correlation parameter, S = Strouhal number, K_{ao} = aeroelastic parameter. Suggested values for $\overline{C_L^2}^{\,1/2}$, L, S, and K_{ao} are:

$$\overline{C_L^2}^{\,1/2} \simeq \begin{cases} 0.45 & Re \lesssim 2x10^5 \quad (5.1.9a) \\ 0.14 & 2x10^5 \lesssim Re \lesssim 2x10^6 \quad (5.1.9b) \\ 0.20 & 2x10^6 \lesssim Re \quad (5.1.9c) \end{cases}$$

$$L \simeq \begin{cases} 2.5 & Re \lesssim 2x10^5 \quad (5.1.10a) \\ 1.0 & Re \gtrsim 2x10^5 \quad (5.1.10b) \end{cases}$$

$$S \simeq \begin{cases} 0.20 & Re \lesssim 1x10^5 \quad (5.1.11a) \\ 0.20 \text{ to } 0.45 & 2x10^5 \lesssim Re < 2x10^6 \quad (5.1.11b) \\ 0.22 & 2x10^6 \lesssim Re \quad (5.1.11c) \end{cases}$$

$$K_{ao} \simeq \begin{cases} 0.6c & Re \lesssim 10^4 \quad (5.1.12a) \\ 1.1c & 10^4 \lesssim Re \lesssim 10^5 \quad (5.1.12b) \\ 0.6c & 10^5 \lesssim Re \quad (5.1.12c) \end{cases}$$

where c = 1.0 for flow with fully developed turbulence and c $\simeq$ 2.0 for relatively smooth flow such as may occur in special atmospheric conditions at low speeds (e.g., at mean hourly wind speeds at 10 m above ground u(10)<10 m/s).

For <u>tapered structures with circular cross-section</u> [528]:

$$\overline{\xi^2_{oi}(z_{ei})}^{1/2} \simeq \frac{0.016\ \overline{C_L^2}^{1/2}\ L^{1/2}\ \rho D^4(z_{ei})\ y_i(z_{ei})}{\zeta_i^{1/2}\ S^2 M_i\ \beta^{1/2}(z_{ei})} \qquad (5.1.13)$$

$$\beta(z_{ei}) \simeq \frac{0.1D(z_{ei})}{z_{ei}} - \left.\frac{dD(z)}{dz}\right|_{z = z_{ei}} \qquad (5.1.14)$$

$$\zeta_{ai}(z_{ei}) \simeq -\frac{\rho D_o^2}{M_i} \int_o^h K_{ao}\left(\frac{u}{u_{cr}}\right) \left[\frac{D(z)}{D_o}\right]^2 y_i^2(z)dz \qquad (5.1.15)$$

$$u_{cr}(z_{ei}) = \frac{n_i\ D(z_{ei})}{S} \qquad (5.1.16)$$

$$\frac{u}{u_{cr}} = \frac{\ln(z/z_o)}{\ln(z_{ei}/z_o)} \qquad (5.1.17)$$

$$K_{ao}\left(\frac{u}{u_{cr}}\right) \simeq \begin{cases} 0 & u/u_{cr} < 0.85 & (5.1.18a) \\ a_t\left(3.5\frac{u}{u_{cr}} - 2.95\right) & 0.85 \leq u/u_{cr} < 1.0 & (5.1.18b) \\ 0.55\ a_t & 1.0 \leq u/u_{cr} < 1.1 & (5.1.18c) \\ a_t\left(2.75 - 2\frac{u}{u_{cr}}\right) & 1.1 \leq u/u_{cr} < 1.3 & (5.1.18d) \\ a_t\left(0.46 - 0.25\frac{u}{u_{cr}}\right) & 1.3 \leq u/u_{cr} < 1.84 & (5.1.18e) \\ 0 & 1.84 \leq u/u_{cr} & (5.1.18f) \end{cases}$$

In eqs. 5.1.18

$$a_t = 1.1\ a_1\ a_2 \qquad (5.1.19)$$

$$a_1 \simeq \begin{cases} 1.0 & Re < 10^4 & (5.1.20a) \\ 1.8 & 10^4 \leq Re < 10^5 & (5.1.20b) \\ 1.0 & 10^5 \leq Re & (5.1.20c) \end{cases}$$

$$a_2 \simeq \begin{cases} 2.0 & u(10m) \lesssim 10m/s & (5.1.21a) \\ 1.0 & u(10m) \gtrsim 10m/s & (5.1.21b) \end{cases}$$

The parameters $\overline{C_L^2}^{1/2}$, and S have the values given in equations 5.1.9 to 5.1.11. To determine the maximum response it is necessary to calculate the value of $\xi_i^2(z_{ei})$ for a sufficiently large number of elevations $0 < z_{ei} < h$.

As pointed out in reference 529, if the structure is very lightly tapered (i.e., if $dD(z)/dz|_{z=z_{ei}}$ and, therefore $\beta(z_{ei})$ is small), then the approximations on which Eq. 5.1.13 is based are no longer valid and equation 5.1.13 ceases to be applicable. In that case, the chimney is assumed to behave as if it had a constant outside diameter D equal to the average diameter of its top third [528], and equations 5.1.7 and 5.1.8 are applied with the same values of the parameters S, $\overline{C_L^2}^{1/2}$, and L as those used in equation 5.1.13.

In practice it is therefore necessary to calculate both the value of the response yielded by equations 5.1.13 and 5.1.15 and the value yielded by equations 5.1.7 and 5.1.8. The response assumed for structural design purposes is the <u>smaller</u> of these two values.

For numerical examples, see reference 111 (Chapter 10).

REFERENCES

501. R. I. Basu, Across-Wind Response of Slender Structures of Circular Cross-Section to Atmospheric Turbulence, Vol. 1, Research Report BLWT-2-1983, The University of Western Ontario, Faculty of Engineering Science, London, Ontrio, Canada, 1983.

502. Bishop, R. E. D. and Hassan, A. Y., "The Lift and Drag Forces on an Oscillating Cylinder," Proceedings of the Royal Society, Series A, Vol. 277, 1964, pp. 32-75.

503. Hartlen, R. T. and Currie, I. G., "Lift Oscillator Model of Vortex-Induced Vibration," Journal of the Engineering Mechanics Division, ASCE, October 1979, pp. 577-591.

504. Skop, R. A. and Griffin, O. M., "On a Theory for the Vortex-Excited Oscillations of Flexible Cylindrical Structures," Journal of Sound and Vibration, Vol. 41, No. 3, 1975, pp. 263-274.

505. Dowell, E. H., "Nonlinear Elasticity," Proceedings, Engineering Foundation Conference in Nonlinear Dynamics, Monterey, California, December 1979 (to be published by SIAM).

506. Griffin, O. M. and Koopman, G. H., "The Vortex-Excited Lift and Reaction Forces on Resonantly Vibrting Cylinders," Journal of Sound and Vibration, Vol. 54, pp. 435-448, 1977.

507. Griffin, O. M. "Vortex-Excited Cross Flow Vibrations of a Single Cylindrical Tube," in Flow-Induced Vibrations, S. S. Chen and M. D. Bernstein (eds.), ASME, New York, pp. 1-10, 1979.

508. Sarpkaya, T., "Transverse Oscillation of a Circular Cylinder in Uniform Flow," Proceedings, ASCE,Journal of Waterways, Ports, Coastal, and Ocean Division, Vol. 104, pp. 275-290, 1978.

509. Jones, G. W., Cincotta, J. J., and Walker, R. W., Aerodynamic Forces on a Stationary and Oscillatory Cylinder at High Reynolds Numbers, NASA Technical Report No. TRR-300, National Aeronautics and Space Administration, Washington, DC, 1969.

510. Toebes, G. H., "Fluidelastic Features of Flow Around Cylinders," Proceedings, International Research Seminar on Wind Effects on Buildings and Structures, Ottawa, Canada, 1967, Vol. 2, University of Toronto Press, Totonto, 1968, pp. 323-334.

511. Ramberg, S. E. and Griffin, O. M., "Vortex Formation in the Wake of a Vibrating, Flexible Cable," Journal of Fluids Engineering, December 1979, pp. 317-322.

512. Ramberg, S. E. and Griffin, O. M., "Velocity Correlations and Vortex Spacing in the Wake of a Vibrating Cable," Journal of Fluids Engineering, March 1976, pp. 10-18.

513. Sallet, D. W., "The Drag and Oscillating Transverse Force on Vibrating Cylinders Due to Steady Fluid Flow," Ingenieur-Archiv, Vol. 44, 1975, pp. 113-122.

514. Bearman, P. W. and Currie, I. G., "Pressure Fluctuation Measurements on an Oscillating Circular Cylinder," Journal of Fluid Mechanics, Vol. 91, pp. 661-678, 1979.

515. Griffin, O. M., "A Universal Strouhal Number for the "Locking-on" of Vortex Shedding to the Vibrations of Bluff Bodies," Journal of Fluid Mechanics, Vol. 85, pp. 591-606, 1978.

516. Griffin, O. M., "Universal Similarity in the Wakes of Stationary and Vibrating Bluff Bodies," in Proceedings, Fifth Internatioal Conference on Wind Engineering, Vol. 1, Pergamon Press, N.Y., 1980.

517. Blevins, R. D. and Burton, T. E., "Fluid Forces Induced by Vortex Shedding," Journal of Fluids Engineering, March 1976, pp. 19-26.

518. Sarpkaya, T., "Vortex-Induced Oscillations, A Selective Review," Transactions, ASME, Journal of Applied Mechanics, Vol. 46, pp. 241-258, 1979.

519. Griffin, O. M., Vortex Shedding from Stationary and Vibrating Bluff Bodies in a Shear Flow, NRL Memorandum Report 4287, Naval Research Laboratory, Washington, DC, August 1980.

520. Rumman, W. S., "Basic Structural Design of Concrete Chimneys," Journal of the Power Division, ASCE, June 1970, pp. 309-318.

521. Maugh, L. C. and Rumman, W. W., "Dynamic Design of Reinforced Concrete Chimneys," Journal of the American Concrete Institute, September 1967.

522. Pinfold, G. M., Reinforced Concrete Chimneys and Towers, Viewpoint Publications, Scholium International, Inc., Flushing, N.Y., 1975.

523. Scruton, C.S. and Flint, A. R., "Wind Excited Oscillations of Structures," Proceedings of the Institution of Civil Engineers, April 1964, pp. 673-702.

524. Vickery, B. J. and Clark, A. W., Lift or Across-Wind Response of Tapered Stacks, Journal of the Structural Division, ASCE, January 1972, pp. 1-20.

525. Davenport, A. G. and Novak, M., "Vibration of Structures Induced by Wind," Chapter 29-II in Shock and Vibration Handbook, 2nd edition, Harris, C. M. and Crede, C. E., eds. McGraw-Hill, New York, 1976.

526. Vickery, B. J. and Basu, R. I., "Across-Wind Vibrations of Structures of Circular Cross-Section, Part I, Development of a Two-Dimensional Model for Two-Dimensional Conditions," Journal of Wind Engineering and Industrial Aerodynamics, 12 (1983), 49-73.

527. Basu, R. I. and Vickery, B. J., "Across-Wind Vibrations of Structures of Circular Cross-Section, Part 2, Development of a Mathematical Model for Full-Scale Applications," Journal of Wind Engineering and Industrial Aerodynamics, 12 (1983), 75-97.

528. Vickery, B. J. and Basu, R. I., "Simplified Approaches to the Evaluation of the Across-Wind Response of Chimneys," Proceedings, 6th International Conference on Wind Engineering, March 1983, Gold Coast, Australia, Journal of Wind Engineering and Industrial Aerodynamics, Vol. 14, 1983, pp. 153-166.

529. Vickery, B. J., "The Aeroelastic Modeling of Chimneys and Towers, Proceedings, Wind Tunnel Modeling for Civil Engineering Applications, Gaithersburg, MD, April 1983, Cambridge University Press, Cambridge - New York, 1982.

6. TRUSSED FRAMEWORKS

The purpose of this chapter is to provide information and references to literature on drag coefficients for trussed frameworks, including frameworks consisting of: a single truss; a pair of parallel trusses; a multiple array of parallel trusses (Sections 6.1, 6.2, and 6.3, respectively). In addition, a brief review is presented of provisions pertaining to trussed frameworks, as set forth in various standards, codes, and design guides (Section 6.4).

For any given wind environment, the principal factors that determine the wind load acting on a trussed framework are: the aspect ratio (i.e., the ratio of the length to the width of the framework), λ ; the solidity ratio (i.e., the ratio of effective to gross area of the framework face), ϕ; the shielding of portions of the framework by other portions located up wind; the shape of the members; the wind direction; the wind gustiness; the dynamic behavior of the framework under fluctuating wind loads; and, in the case of members with rounded cross section, the Reynolds number and the roughness of the member surface. In the material that follows, the drag coefficients are referred to, and should be used in conjunction with, the effective area of the framework face, A_f.

6.1 SINGLE TRUSSES

Figures 6.1.1 and 6.1.2 summarize measurements of the drag coefficient for a single truss normal to the wind, $C_d^{(1)}$, reported in the early 1930's by Flachsbart [601,602,603]. Also shown in figure 6.1.1 are results reported in 1979 [604] for trusses composed of flat-sided members and of circular members in subcritical flows. Figure 6.1.3 [605] summarizes results of research on trusses with circular members conducted in the subsonic wind tunnel at Porz-Wahn, West Germany [606] and in the compressed air tunnel of the National Physical Laboratory, U.K. [607]. Note that for Re $< 10^5$ the drag coefficients of figure 6.1.3 [605] are somewhat larger than those indicated for circular members in figure 6.1.1 [604].

For single trusses, the wind direction normal to the truss causes the highest drag force.

6.2 PAIRS OF TRUSSES

Let $C_d^{(2)}$ denote the drag coefficient for a pair of parallel trusses normal to the wind, and let the contributions to $C_d^{(2)}$ due to the front and rear truss be denoted by $C_d^{(1)}\eta_1$, and $C_d^{(1)}\eta_2$, respectively. Measured values of η_1 and η_2 reported in reference 604 for trusses with infinite aspect ratio are given in figures 6.2.1 (trusses with flat sided members) and 6.2.2 (trusses with circular members in subcritical flow). For the sake of simplification $C_2^{(2)}$ is expressed in the form:

$$C_d^{(2)} = C_d^{(1)}\ (1 + \eta_e^{(2)}) \qquad (6.2.1)$$

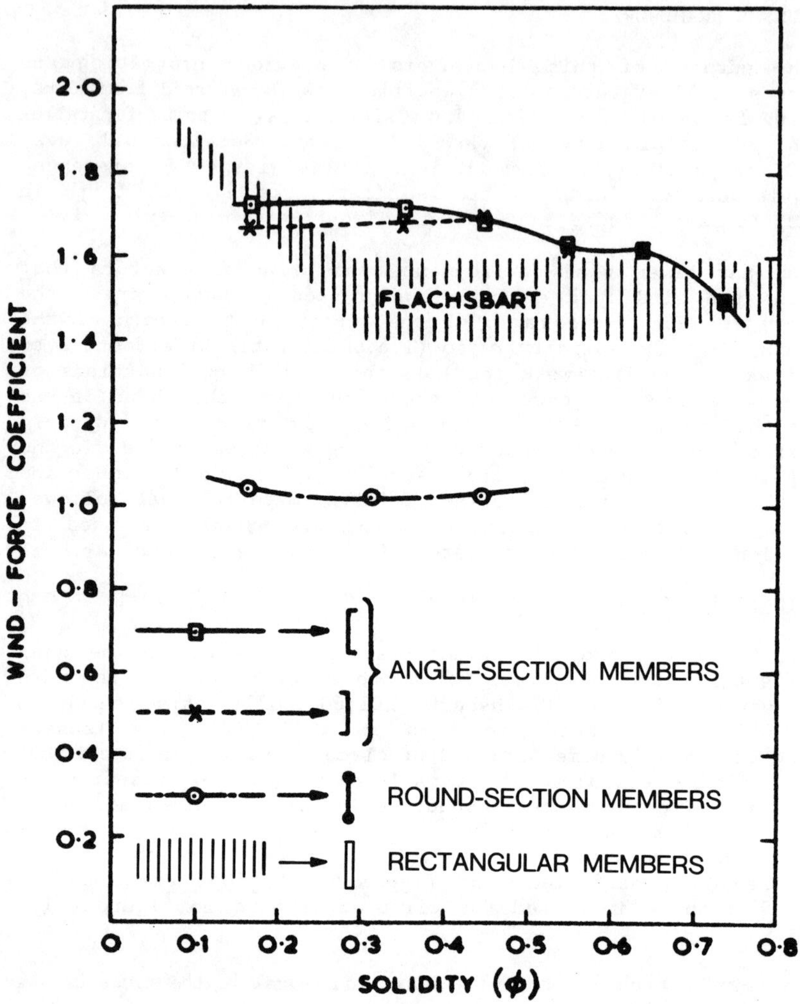

Figure 6.1.1 Drag Coefficients for Single Truss Normal to the Wind (Aspect Ratio $\lambda = \infty$) [603,604].

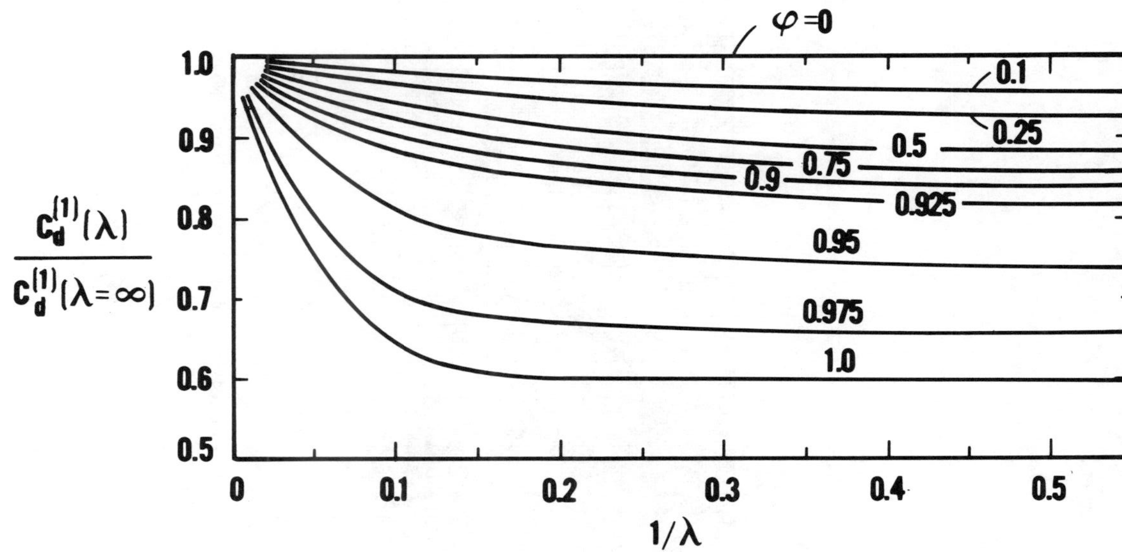

Figure 6.1.2 Reduction of Drag Coefficient for Single Truss Normal to the Wind as a Function of Aspect Ratio [603].

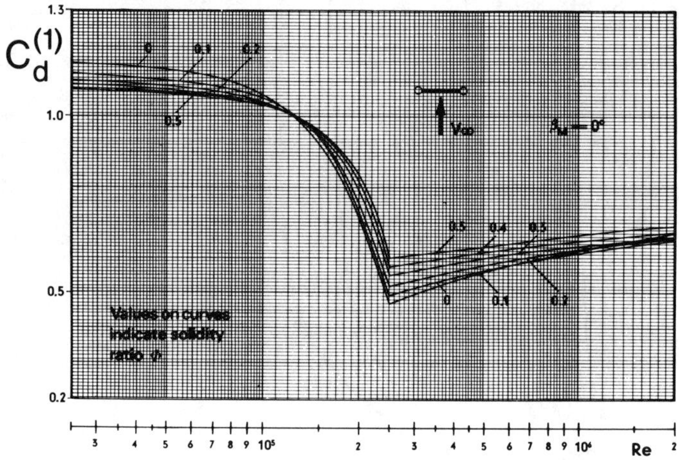

Figure 6.1.3 Drag Coefficients for Single Truss Normal to the Wind (Round-Section Members) [605].

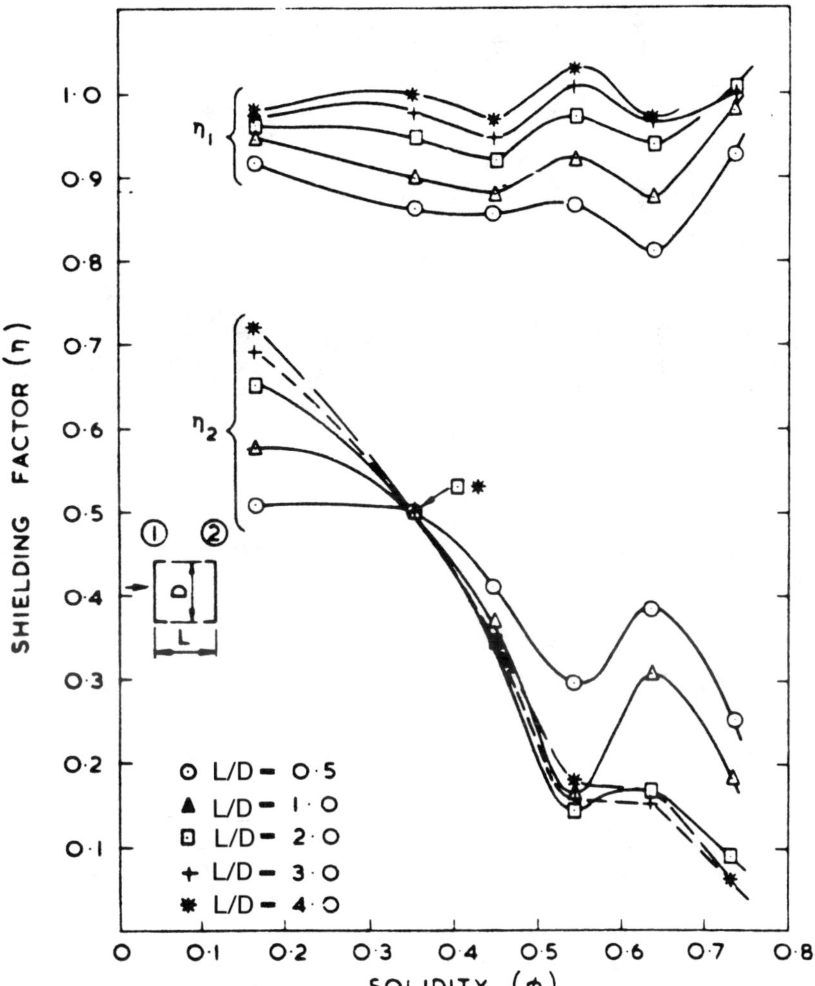

Figure 6.2.1 Shielding Factors for Pair of Trusses Normal to the Wind, Angle-Section Members [604].

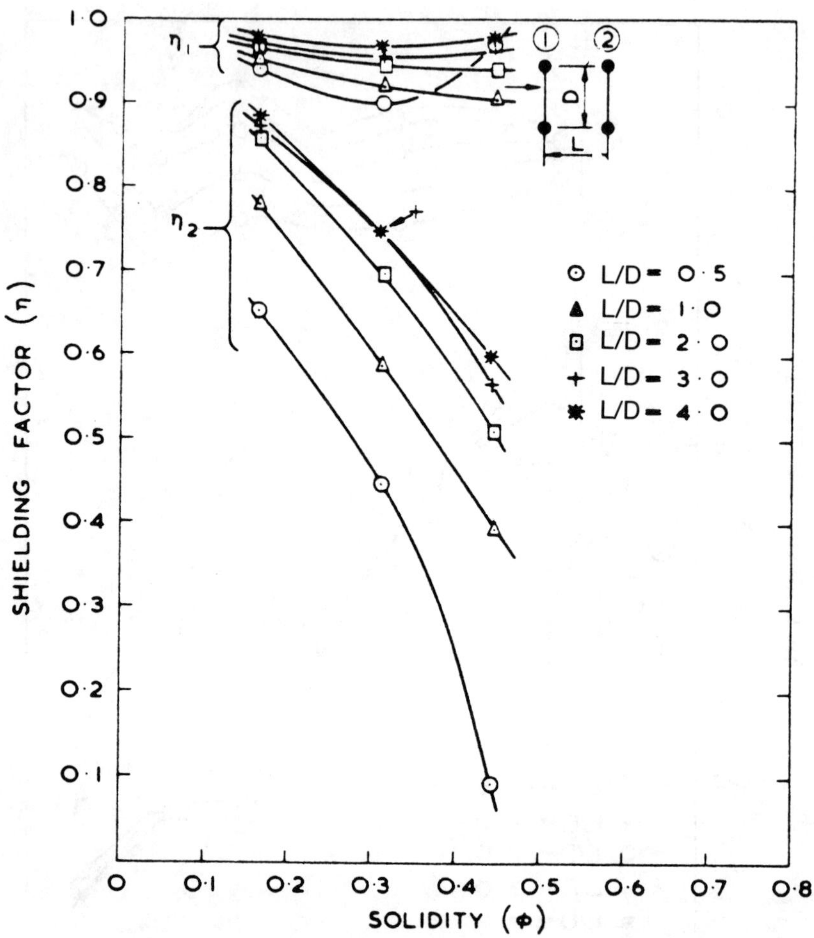

Figure 6.2.2 Shielding Factors for Pair of Trusses Normal to the Wind, Round-Section Members [604].

According to Flachsbart [603, 604, 608]:

$$\eta_e^{(2)} \simeq 1.15[1 - 1.45\ \phi\ (L/D)^{-0.25}] \qquad 0.1 < \phi < 0.6 \qquad (6.2.2)$$

where L = depth (along-wind dimension) and D =width of trussed framework (see figure 6.2.1). However, on the basis of wind tunnel measurements, reference 604 proposes the expression

$$\eta_e^{(2)} = 1 - \phi^{0.45}(L/D)^{\phi-0.5} \qquad 0 < \phi < 0.5 \qquad (6.2.3)$$

Measured values of $\eta_e^{(2)}$ recently reported in reference 601 for trusses with flat-sided members (aspect ratio λ =8) are shown in figure 6.2.3.

Table 6.2.1 includes values of $C_d^{(2)}$ calculated for trusses with flat-sided members and aspect ratio λ =8 based upon: (1) Flachsbart's values for $C_d^{(1)}$ (fig. 6.1.1) and Flachsbart's model for $\eta_e^{(2)}$ (eq. 6.2.2); (2) the measured values of $C_d^{(1)}$, η_1, and η_2 from reference 604 (fig. 6.1.1 and 6.2.1); (3) the measured values of $C_d^{(1)}$ and the model for $\eta_e^{(2)}$ from reference 604 (fig. 6.1.1 and eq. 6.2.3); and (4) Flachsbart's values for $C_d^{(1)}$ (fig. 6.1.1) and the measured values of $\eta_e^{(2)}$ reported in reference 601 (fig. 6.2.3). It is seen that the drag coefficients based on reference 601 are in good agreement with those based on Flachsbart's work for all values ϕ and L/D of table 6.2.1.

Note, on the other hand, that for ϕ = 0.136 the drag coefficients based on reference 604 are lower by about 25 percent than those based on Flachsbart's work and on reference 601.

A similar comparison is of interest for frameworks composed of circular members. Figure 6.2.4, applicable to a square framework, is excerpted from reference 605 and is based on tests reported in references 606 and 607. Note that for ϕ = 0.16, ϕ = 0.35 and ϕ = 0.45, the values of $C_d^{(2)}$ given in reference 604 (figs. 6.1.1 and 6.2.2) are 1.90, 1.55, and 1.40, respectively, whereas the values of reference 605 (figure 6.2.4) are approximately 1.95, 1.55, and 1.50, respectively. The differences between the two sets of results are 8 percent or less.

Information on the dependence of the wind force upon the angle between the mean wind velocity and the normal to the face of the truss is given in references 601 through 603, and 606 through 610. It is noted that for ϕ > 0.1 or so, drag forces corresponding to skew directions can be larger than those caused by winds normal to the parallel trusses.

6.3 MULTIPLE-FRAME ARRAYS

Let the drag coefficient, $C_d^{(n)}$, for an array of n frames with infinite aspect ratio normal to the wind be expressed in the form

$$C_d^{(n)} = C_d^{(1)}\ (\eta_1 + \eta_2 + \ldots + \eta_n) \qquad (6.3.1)$$

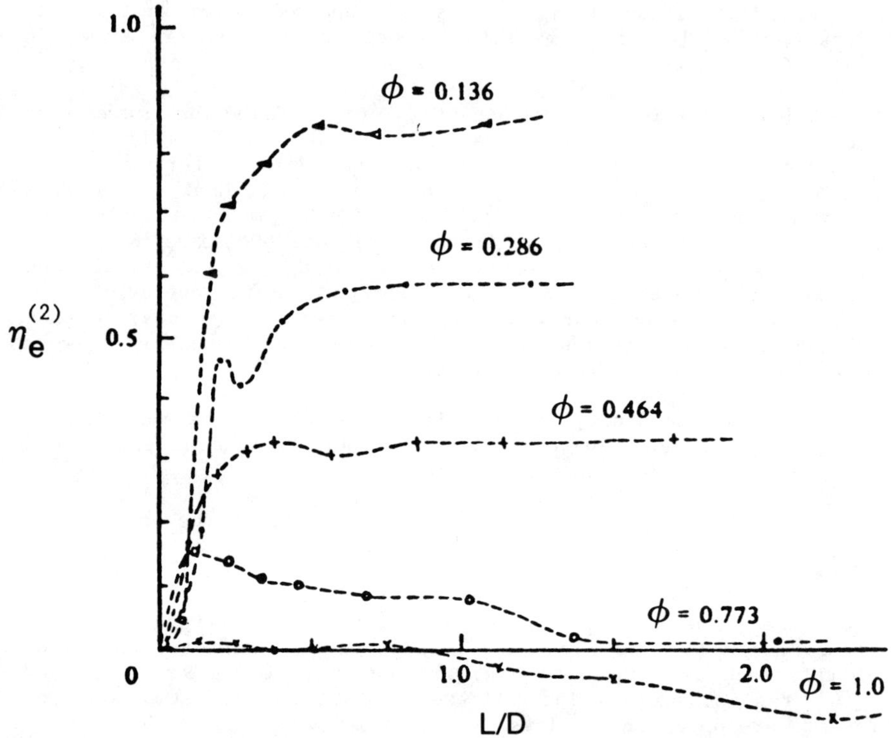

Figure 6.2.3 Factor $\eta_e^{(2)}$ [601].

Table 6.2.1. Values $C_d^{(2)}$

ϕ	Case	L/D = 0.5	L/D = 1.0
0.136	(1)[a]	3.20	3.25
	(2)[b]	2.40	2.50
	(3)[c]	2.50	2.70
	(4)[d]	3.25	3.25
0.286	(1)[a]	2.20	2.30
	(2)[b]	2.20	2.35
	(3)[c]	2.20	2.30
	(4)[d]	2.30	2.35
0.464	(1)[a]	1.70	1.90
	(2)[b]	1.95	1.95
	(3)[c]	1.90	1.95
	(4)[d]	1.90	1.90

[a] After Flachsbart (eq. 6.2.2).

[b] After reference 604 (measured values).

[c] After reference 604 (eq. 6.2.3).

[d] After reference 601.

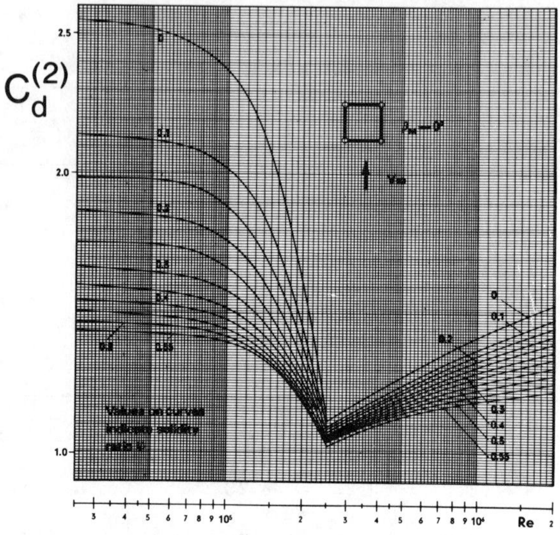

Figure 6.2.4 Drag Coefficients for Trussed Framework Normal to the Wind, Round Section Members [605].

Shielding factors $\eta_1, \eta_2, \ldots \eta_n$ obtained by measurement in reference 604 are shown in figures 6.3.1 and 6.3.2 for frameworks with flat-sided members and infinite aspect ratio. Let $C_d^{(n)}$ be expressed in the alternative form

$$C_d^{(n)} \simeq C_d^{(1)} \, (1 + \eta_e^{(n)}) \tag{6.3.2}$$

Reference 604 proposes the following model for frameworks with infinite aspect ratio normal to the wind

$$\eta_e^{(n)} = n - 1 - \phi^{0.45}(L/D)^{(\phi - 0.5)} \, (n - 1)^{1.1} \qquad 0 < \phi < 0.5 \tag{6.3.3}$$

where n = number of frames, L = along-wind distance between two successive trusses, and D = width of framework in a direction normal to L. According to reference 604, equations 6.3.2 and 6.3.3 may be used for frameworks with either flat-sided or circular members.

For large values of ϕ (e.g., $\phi > 0.35$), the measured data and the model for $\eta_e^{(n)}$ of reference 604 yield values of $C_d^{(n)}$ that are comparable with those reported in reference 601. However, for smaller values of ϕ (e.g., $\phi = 0.16$), the values of reference 604 are smaller by about 15 to 20 percent than those of reference 601.

Reference 601 includes data concerning the drag force on multiple-frame arrays whose trusses are not normal to the mean wind. For a truss with n = 5 and L/D = 1, the ratio of the drag corresponding to the most unfavorable wind direction to the drag induced by wind normal to the trusses is about 1.3 for $\phi = 0$ and about 2.5 for $\phi = 0.5$. Results of tests conducted on various types of trussed frameworks with cylindrical members used for communications by the British Post Office are reported in references 611, 612, 613, and 614. Reference 612, which is devoted to the study of the effects of turbulence upon the mean aerodynamic forces, concluded that these effects were negligible in the case of square trussed towers with circular members. This result is of interest because previous tests establishing the mean drag coefficients are not affected by turbulence had been reported only for trusses with flat-sided members [603, p. 75].

6.4 SQUARE AND TRIANGULAR FREE-STANDING TOWERS

In view of the importance of square and triangular towers, it is appropriate to review briefly provisions concerning their wind loading, as set forth in various standards, codes, and design guides. In this section it will be assumed that all the faces of the towers are identical.

The wind force per unit length acting on a tower at elevation z, F(z), may be expressed as follows:

$$F(z) = \tfrac{1}{2}\rho V^2(z) \, C_d^{(1)} c_s c_m c_\theta G A_f \tag{6.4.1}$$

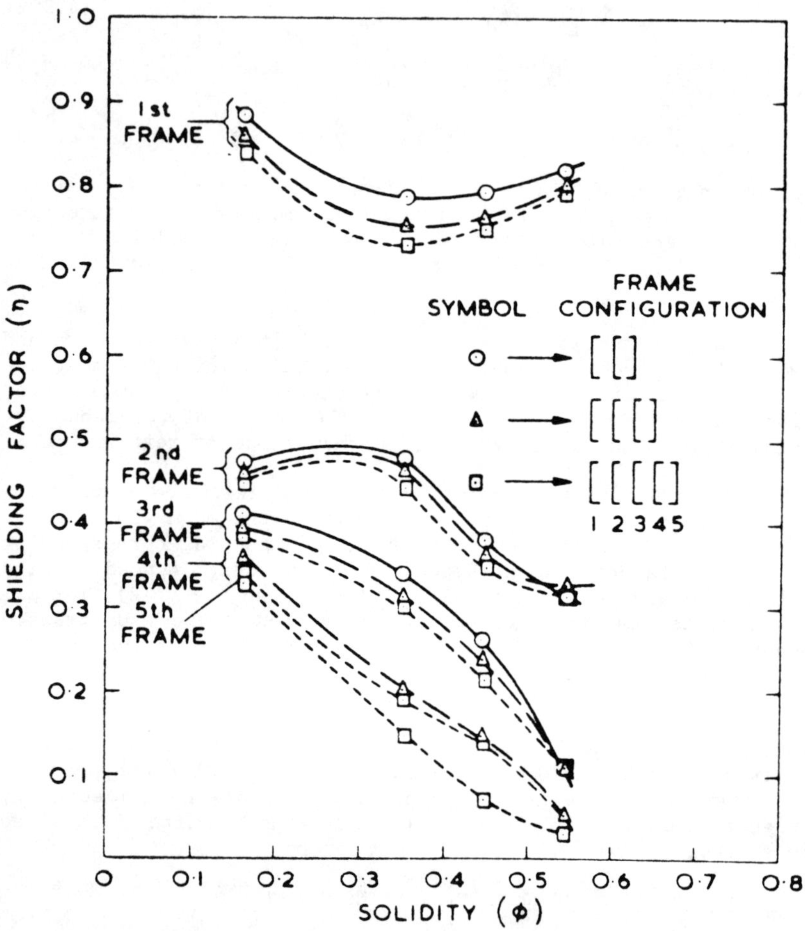

Figure 6.3.1 Shielding Factors for Multiple-Frame Arrays, L/D = 0.5, Angle-Section Members [604].

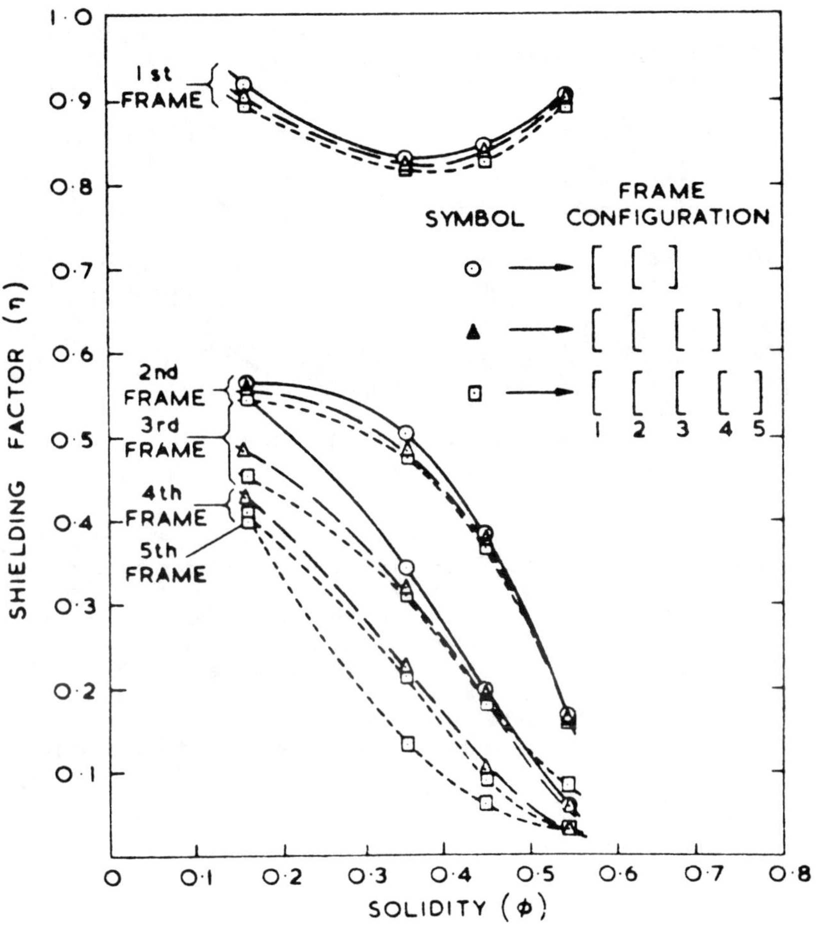

Figure 6.3.2 Shielding Factors for Multiple-Frame Arrays, L/D = 0.5, Round-Section Members [604].

where

ρ = air density;

$V(z)$ = wind speed at elevation z;

$C_d^{(1)}$ = nominal force coefficient characterizing mean aerodynamic force that would act on a plane truss normal to the wind and identical to a face of the tower;

c_s = coefficient dependent upon the configuration of the tower (i.e., whether square or triangular) and its solidity ratio (i.e., ratio of effective to gross area of a face), ϕ ;

c_m = ratio of force acting on tower composed of circular (cylindrical) members to force acting on tower with flat-sided members;

c_θ = ratio of force corresponding to wind acting on corner to force induced by wind normal to a face;

G = gust factor accounting for effect of wind speed fluctuations; and

A_f = effective area of a tower face.

Wherever possible, provisions in codes, standards, and design guides will be interpreted in terms of the coefficients of eq. 6.4.1. Note that such interpretations are not necessarily unique. For example, if a provision specifies a coefficient equal to the product $C_d^{(1)}c_s$, various choices of $C_d^{(1)}$ are possible, which will determine the corresponding value of c_s. Nonetheless, the format offered by eq. 6.4.1 facilitates the comparative assessment of code provisions and of the validity of various assumptions implicit therein.

6.4.1 American National Standards ANSI A58.1 [119]

The wind speed V(z) is expressed in fastest-mile wind speeds (mph) and is given by the relation

$$V(z) = V_{ref}\, b\left(\frac{z}{z_{ref}}\right)^{\alpha} \qquad (6.4.2)$$

where V_{ref} = fastest-mile wind speed over open terrain at the reference elevation z_{ref}, b and α = coefficient and exponent depending upon terrain category (see table 1.2.3 and eq. 1.2.8), and z_{ref} = 33 ft (10 m). The gust factor G is tabulated in reference 119 for towers with fundamental frequency of vibration $n_1 > 1$ Hz. For example, over open terrain G varies from 1.32 for heights of 15 ft or less to 1.00 for heights of approximately 600 ft or more. For towers with $n_1 < 1$ Hz, reference 119 requires that G be calculated by a rational method. (The procedures of chapter 4, with appropriate modifications to account for the variation of tower width with height, may be employed for this purpose.) The force coefficient $C_d^{(1)}$ is 2.00. The coefficient c_s is shown in table 6.4.1. For towers with circular members c_m depends upon ϕ as shown in table 6.4.2.

Table 6.4.1 Coefficient c_s Implicit in Reference 119

ϕ	Square Towers	Triangular Towers
< 0.025	2.00	1.8
0.025-0.44	$2.050-2.6\phi$	$1.85 - 2.25\phi$
0.45 -0.69	0.90	0.85
0.70 -1.00	$0.650 +0.350\phi$	$0.5(1 + \phi)$

Table 6.4.2 Coefficient c_m Implicit in Reference 119

Solidity ratio, ϕ	c_m
Less than 0.30	2/3
0.30 to 0.79	$0.66\phi + 0.47$
0.80 to 1.00	1.00

For square towers $c_\theta = 1+0.75\phi$ (ϕ <0.5). For triangular towers $c_\theta = 1.00$.

Note that in the 1972 version of the ANSI A58.1 Standard the values of table 6.4.2 are specified to be valid only for $d\sqrt{q} < 2.5$, where d = typical member diameter in feet, and q = dynamic pressure in psf based on the fastest-mile wind speed.

This condition is equivalent to $Re < 2 \times 10^5$ (where the speed used to calculate Re is expressed in terms of the fastest-mile wind). No allowance is made in reference 119 (as well as in its 1972 version) for the reduction of the drag force on trussed towers that occurs at higher Reynolds numbers, even though such a reduction is provided for in the case of plane trusses.

In reference 119 ice loading is classified as a live load, and no guidance is provided on ice thicknesses.

The provisions of reference 119 are largely based upon the interpretation of Flachsbart's [602,603] and Katzmayr's [609] experimental results by Pagon [615], Cohen and Perrin [616], Cohen [617], and Cohen, Vellozzi, and Suh [618].

6.4.2 ASCE Guide for Design of Steel Transmission Towers [619].

Reference 619 specifies for steel transmission towers three separate loading conditions and corresponding design criteria:

(1) Constant wind velocity, V_c, in a direction normal to the transmission line, acting on the tower assumed to be free of ice and on the wires assumed to be covered by a layer of ice with radial thickness, R. The United States is divided into three zones for which V_c = 39.2 mph, R = 0.5 in; V_c = 39.2 mph, R = 0.25 in; and V_c = 58.8 mph, R = 0, respectively. The completed structure shall support without permanent set in any member 2.54 times the load induced by loading condition (1).

(2) 50-yr fastest-mile wind velocity as specified in the 1972 ANSI A58.1 Standard, acting normal to the transmission lines in the absence of ice. The completed structure shall support without permanent set 1.00 times the load induced by loading condition (2).

(3) 50-yr fastest-mile wind velocity as specified in the 1972 ANSI A58.1 Standard, acting from any direction on the tower standing without wires in the absence of ice. The structure shall support without permanent set 1.3 times the load induced by loading condition (3).

An additional loading condition is specified for the case where a change in the direction of wires occurs. For loading conditions (2) and (3) the wind speed may be assumed to be uniform with height provided that the height of the tower is less 200 ft. For higher towers more elaborate assumptions on wind speed variation with height are recommended, e.g., based on reference 101.

Note that for loading condition (1)--the only loading condition involving ice loads--the wind speeds are less than those specified by reference 119. This is justified for the zones in which ice is taken into consideration by the fact the probability of occurrence of extreme winds is smaller in the presence than in the absence of ice.

To estimate wind loads eq. 6.4.1 is used, where for condition (1) $V(z) = V_c$. For conditions (2) and (3), V(z) = 50-yr fastest mile wind speed at elevation z over open terrain; however, for towers with height less than 200 ft, ref. 619 does not clearly specify the elevation of the constant fastest-mile wind speed, i.e., whether the fastest-mile at 30 ft, at the top of the tower, or any other elevation shall be used . The gust factor G is 1.0 for towers supporting wires, and 1.3 for towers standing without wires. The force coefficient is $C_d^{(1)} = 2.00$. Implicit in reference 619 are, for both square and triangular towers, the coefficient c_s =

1.64 for $\phi \leq 0.5$ and $c_s = 0.82/\phi$ for $\phi > 0.5$. For towers with circular members $c_m = 0.61$ for all Reynolds numbers. For all towers $c_\theta = 1.00$.

6.4.3 Electronic Industries Association EIA-RS-222-C Standard [620].

Reference 620 requires that ice be considered in design but does not include ice thickness information or specifications. For towers with heights less than 300 ft or more than 650 ft wind loads are specified as uniform throughout the tower height. For towers with heights between 300 ft and 650 ft two zones of uniform pressure are specified, covering the lower 300 ft and the portion above the 300 ft level, respectively. The United States is divided into three wind zones corresponding, roughly, to the 70 mph, 80 mph, and 90 mph 50-yr fastest-mile wind zones defined in reference 119 for open terrain. The respective wind pressures are shown in table 6.4.3. Note that design in accordance with table 6.4.3 may lead to significant differences (of more than 40 percent) between loadings on towers slightly lower than 650 ft on the one hand, and slightly higher than 650 ft on the other hand.

Table 6.4.3 Wind Pressure (in psf) on Flat-sided Members Specified in EIA Standard

Tower Height, h (ft)	Height above ground (ft)	Wind Loading Zone A	B	C
≤ 650	< 300	30	40	50
	300 - 650	35	48	60
> 650	0 - h	50	65	85

It is indicated in reference 620 that wind pressures may be calculated by using the expression

$$p = 0.004\ V^2 \tag{6.4.3}$$

where the coefficient 0.004 is the product of the dynamic pressure by a gust factor and a drag coefficient for flat surfaces. The pressures of table 6.4.3 may than be interpreted as corresponding to 50-yr fastest-mile wind speeds over open terrain at some elevation below the tower top (e.g., two-thirds the tower height, h, or so), with $G \simeq 2.00$ for $h > 600$ ft. For square towers $c_s = 1.75$ and for triangular towers $c_s = 1.50$. For towers with circular members $c_m = 2/3$ regardless of Reynolds number. For all towers $c_\theta = 1.00$.

With regard to forces induced on surfaces covered with ice, reference 620 specifies that the portion of the wind force acting on the added area created by the accumulated ice should be affected by the factor $c_m = 2/3$.

6.4.4 Naval Facilities Engineering Command NAVFAC DM-2 Design Manual [621]

NAVFAC provisions apply to towers with heights below 300 ft. It is assumed, in all cases, that open terrain conditions prevail. Referring to eqs. 6.4.1 and 6.4.2, V_{ref} = 25-yr peak gust speed at elevation z_{ref} (z_{ref} = 30 ft) and G = 1.00 or, equivalently, V_{ref} = 50-yr fastest-mile wind speed at 33 ft elevation (as in reference 119) and G is equal to the square of the ratio of the 25-yr peak gust at 30 ft elevation to the 50-yr fastest-mile at 33 ft elevation, i.e., $G \simeq 1.30$. The NAVFAC provisions may be interpreted as implying: $C_d^{(1)} = 2.00$; for square towers a factor c_s varying from 1.60 (for $\phi = 0.08$) to 0.92 (for $\phi = 1.0$), and for triangular towers $c_s \simeq 1.145$ for all solidity ratios ϕ ; for towers with circular members $c_m = 2/3$ regardless of Reynolds number; for square towers $c_\theta = 1.2$ and for triangular towers $c_\theta = 0.846$. Ice thicknesses specified in reference 621 are the same as those of reference 619, except that the provisions of reference 621 may be interpreted as also requiring consideration of ice accumulation on the structures themselves, rather than just on the transmission lines.

6.4.5 Canadian Standards Association (CSA) S37-1976 Standard [622]

The wind forces are calculated using eq. 6.4.1, where V = hourly mean speed with a 30-yr mean recurrence interval and G =2.00 regardless of tower height (or, equivalently, V =fastest-mile wind speed with a 50-yr mean recurrence interval and $G \simeq 1.20$); $C_d^{(1)}$ = 2.0 for all towers; $c_s = 1.5$ for square towers and $c_s = 1.125$ for triangular towers; for towers with flat sided members $c_m = 1.0 - R/1.5$ ($R \leq 0.3$) and $c_m = 0.8$ ($R > 0.3$), where R = ratio of radial ice thickness to member width; for towers with circular members $c_m = 2/3$; $c_\theta = 1.0$ for all towers.

6.4.6 British Code of Practice CP3: Chapter V : Part 2: 1972 [623]

Wind speeds are expressed in terms of 5 sec averages for towers with height $h \leq 50$ m and in terms of 15 sec averages for towers with heights $h > 50$ m. If in eq. 6.4.1 the wind speeds being used are fastest-miles, the gust factors implicit in reference 593 are, for open terrain, $G \sim 1.30$ for $h \leq 50$ m and $G \simeq 1.15$ for $h > 50$ m. Force coefficients consistent with the provisions of reference 623 are:

$C_d^{(1)} = 2.0$

square towers with flat-sided members -- $c_s = 1.9$ ($\phi = 0.1$) to $c_s \simeq 1.05$ ($\phi = 0.5$)

triangular towers with flat-sided members $c_s = 1.55$ ($\phi = 0.1$) to $c_s = 0.75$ ($\phi = 0.5$)

square towers with cylindrical members -- $c_s = 1.9$ ($\phi = 0.05$) to c_s 1.11 ($\phi = 0.5$, low Reynolds numbers) or $c_s \sim 2.3$ ($\phi = 0.5$, high Reynolds number)

triangular towers with cylindrical members -- $c_s = 1.43$ ($\phi = 0.05$) to c_s 1.1 ($\phi = 0.5$, low Reynolds numbers) or $c_s = 2.0$ ($\phi = 0.5$, high Reynolds numbers)

towers with cylindrical members -- $c_m \sim 0.63$ (low Reynolds numbers) or $c_m \sim 0.3$ (high Reynolds numbers)

square towers with flat-sided members -- $c_\theta = 1.2$
for square towers with cylindrical members -- c_θ is a function of ϕ and Reynolds number

triangular towers $c_\theta = 1.0$.

It is seen that one major difference between the British Code of Practice and all other standards and guides reviewed in this section is that the former takes into account the dependence of the drag forces on towers with cylindrical members upon Reynolds number. This is done on the basis of the results of references 606 and 607.

6.4.7 CONSTRADO Manual "Wind Forces on Unclad Tubular Structures" [605]

The CONSTRADO Manual presents calculation procedures based upon the results of references 606 and 607. References 605 includes: a simplified and an elaborate procedure applied to structures with circular members; a procedure applicable to structures with square members with rounded corners, and a procedure for calculating the effect of gusset plates and other parts (e.g., ladders). This section will reproduce the simplified procedure given in reference 605 for structures with circular members. It is shown in reference 605 that differences between loads estimated by the simplified and the elaborate procedure given therein can be expected to be approximately 1 percent to 6 percent.

6.4.7.1 Approximate Calculation of Wind Loads on Trussed Towers (after reference 605)

The steps to be followed are:

(1) Calculation of Reynolds number and dynamic pressure;

$$Re = \frac{Vd_G}{\nu} \qquad \text{(see eq. 2.1.6)}$$

$$q = \frac{\rho V^2}{2} \qquad \text{(see eq. 2.1.4)}$$

where d_G = diameter of chord, ν = kinematic viscosity, ρ = density, V = wind speed. The speed V is not defined in reference 605. It is assumed here that V is specified as in reference 623.

(2) Calculation of solidity ratio, ϕ;

$$\phi = \frac{A_f}{A_g} = \frac{\text{effective frontal area}}{\text{gross area}}$$

The formula can also be used:

$$\phi = \frac{2d_G}{b}\left[1 + \frac{n}{2}\left(1 - \frac{d_G}{b}\right)(1 + \tan^2\gamma_B)^{1/2}\right]$$

b = Distance between centerline of two chords (fig. 6.4.1)

d_G = Diameter of chord

d_B = Diameter of bracing member

$n = d_B/d_G$

γ_B = Angle of bracing members to the vertical = $\tan^{-1}\dfrac{b}{\ell_B}$

ℓ_B = Half panel length of tower section.

(3) Force coefficients C_d or C_w are obtained from figures 6.2.2 (where $C_d^{(2)} = C_d$) and 6.4.2 to 6.4.4 [C_d and C_w = force coefficient in the along-wind direction corresponding to case where wind velocity is normal and skewed with respect to a tower face, respectively].

(4) The forces F or W corresponding to the coefficients C_d and C_w are calculated as follows:

$$F = C_d\, q\, A_f \tag{6.4.3a}$$

$$W = C_w\, q\, A_f \tag{6.4.3b}$$

Numerical Example (Fig. 6.4.1). Triangular trussed tower; wind direction along median (fig. 6.4.3). The structure has the following dimensions:

$d_G = 0.0337$ m, $d_B = 0.0214$ m, $b = 0.300$ m, $\ell_B = 0.356$ m

$$\tan\gamma_B = \frac{0.300}{0.356} = 0.8426$$

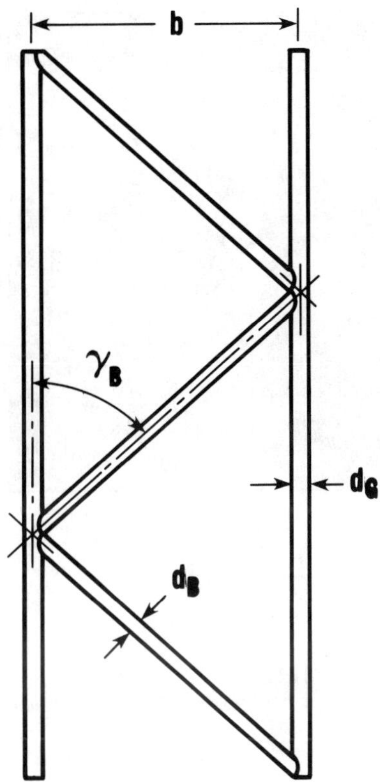

Figure 6.4.1 Truss Elevation and Notations [605].

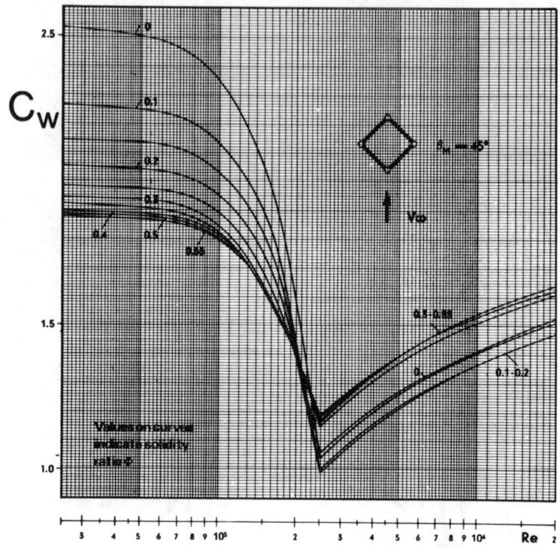

Figure 6.4.2 Drag Coefficient for Wind Acting Along Diagonal, Square Framework with Round-Section Members [605].

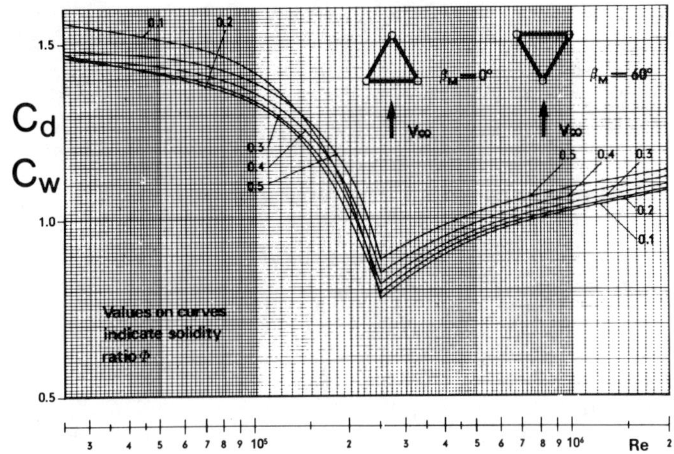

Figure 6.4.3 Drag Coefficients for Wind Acting Along Median, Triangular Framework with Round-Section Members [605].

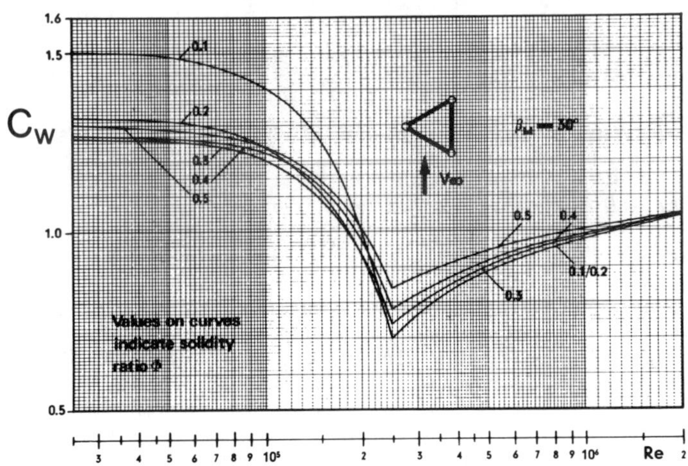

Figure 6.4.4 Drag Coefficients for Trussed Framework, Wind Acting Parallel to a Face, Triangular Framework with Round-Section Members [605].

At a wind velocity V = 30 m/sec the Reynolds number is:

$$Re = \frac{Vd_G}{\nu} \approx 67{,}000 \times 30 \times 0.0337 \qquad \text{(eq. 2.1.6)}$$

$$= 0.68 \times 10^5 \text{ (subcritical)}$$

The solidity ratio ϕ is :

$$\phi = \frac{2 \times 0.0337}{0.300} \left[1 + \frac{0.0214}{2 \times 0.0337} \left(1 - \frac{0.0337}{0.300}\right) \right.$$

$$\left. \times (1 + 0.8426^2)^{1/2}\right]$$

$$\phi = 0.3075$$

With the above Re and ϕ-values, figure 6.4.3 gives:

$$C_d = 1.39 \text{ (subcritical)}$$

The normal force acting on the entire triangular lattice structure is then calculated as follows:

a) $q = 0.612\ V^2 = 551\ N/m^2$ (eq. 2.1.4)

b) The effective frontal area equals the gross area times the solidity ratio:

$A_f = \phi \cdot b \cdot h$ (h = height of tower). It is assumed h = 7.12 m, then

$$A_f = 0.3075 \times 0.3 \times 7.12 = 0.6568\ m^2$$

c) $F = q \cdot C_d \cdot A_f = 551 \times 1.39 \times 0.6568 = 503\ N$

Reference 605 also provides methods for estimating the distribution of the total load among the planes of the trussed tower, and additional wind forces due to gusset plates and other superstructural parts (e.g., ladders).

6.4.8 Summary of Coefficients C_d, G, c_s, c_m, c_θ Implicit in Various Codes, Standards, and Design Guides

Differences in formulation among various expressions for the wind loading on trussed towers, as well as differences in the physical assumptions on which these expressions are based, make it difficult to effect rigorous and exhaustive comparisons among code and standard provisions. As noted previously, this task is facilitated by the interpretation of these provisions in terms of the coefficients of eq. 6.4.1. Approximate values or ranges for these coefficients are listed in table 6.4.4 for a number of typical cases.

Table 6.4.4 Coefficients C_d, G, c_s, c_m, c_θ, Implicit in Various Codes, Standards, and Guides for Typical Design Situations

Document	ANSI A58.1 (Ref. 119)		ASCE Guide (Ref. 619)	EIA (Ref. 620)		NAVFAC (Ref. 621)		CSA Standard (Ref. 622)		BCP (Ref. 623)	
Tower Configuration	Square	Triangular	All Configurations	Square	Triangular	Square	Triangular	Square	Triangular	Square	Triangular
$C_d^{(1)}$	2.0	2.0	2.0	1.4^c–2.0^d	1.4^c–2.0^d	2.0	2.0	2.0	2.0	2.0	2.0
G	1.0 to 1.32^a	1.0 to 1.32^a	1.30^b	1.1^c–1.05^d	1.1^c to 1.05^d	1.3	1.3	1.20^j	1.20^j	1.3^j to 1.15^j	1.3^j to 1.15^j
c_s	2.0 to 1.0	1.8 to 1.0	$1.64(\phi^h < 0.5)$ $0.82(\phi^h > 0.5)$	1.75	1.50	1.6 to 0.92	1.145	1.50	1.125	1.9 to 1.05^k	1.55 to 0.75^k
c_m^e	0.67 to 1.0	0.67 to 1.0	0.61	0.67	0.67	0.67	0.67	0.67^f	0.67^f	See ref. 623.	See ref. 623.
c_θ	$1 + 0.75\phi^h$	1.07	1.0	1.0	1.0	1.2	0.845	1.0	1.0	1.2^k	1.0^k

[a] For towers with fundamental natural frequency of vibration $n_1 > 1$ Hz in open terrain. The lower values of G apply to taller towers.

[b] For towers standing without wires; stresses in tower should not exceed yield stress.

[c] For towers with heights less than 300 ft.

[d] For towers with heights in excess of 600 ft.

[e] For towers with cylindrical members. (For towers with flat-sided members $c_m = 1.00$.)

[f] For flat-sided member covered with ice $c_m = 1.0 - R/1.5$ (if $R \leq 0.3$) and $c_m = 0.8$ (if $R > 0.3$); R ratio of radial ice thickness to member width.

[g] c_m depends upon solidity ratio and Reynolds number.

[h] ϕ = solidity ratio.

[j] Referred to 50-yr fastest-mile speed.

[k] For towers with flat-sided members.

An examination of table 6.4.4 reveals the existence of significant differences among the various provisions (e.g., the ANSI and NAVFAC provisions on triangular towers). Such differences suggest that additional research into the wind loading of trussed towers is warranted.

Comparisons of wind loads inherent in various U.S. standards and guides are given in reference 624.

REFERENCES

601. Georgiou, P. N. and Vickery, B. J., "Wind Loads on Building Frames," Wind Engineering, Proceedings of the Fifth International Conference, Fort Collins, July 1979, Pergamon Press, Oxford-New York, 1980, pp. 421-433.

602. Flachsbart, O., "Winddruck auf vollwändige Bauwerke und Gitterfachwerke," Publications of the International Association of Bridge and Structural Engineering, No. 1, Zurich, 1932, pp. 133-172.

603. Flachsbart, O. and Winter, H., Modellversuche über die Belastung von Gitterfachwerken durch Windkräfte, Der Stahlbau, No. 9, 1934, pp. 65-69, No. 10, 1934, pp. 73-79, and No. 8, 1935, pp. 53-63.

604. Whitbread, R. E., "The Influence of Shielding on the Wind Forces Experienced by Arrays of Lattice Frames," Wind Engineering, Proceedings of the Fifth International Conference, Fort Collins, July 1979, Pergamon Press, Oxford - New York, 1980, pp. 405-420.

605. Wind Forces on Unclad Tubular Structures, H. B. Walker, ed., Constrado Publications 1/75, Constructional Steel Research and Development Organization, Croydon, U.K., 1975.

606. Schulz, G., The Drag of Lattice Structures Constructed from Cylindrical Members (Tubes) and its Calculation, CIDECT Report No. 69/29, Düsseldorf, West Germany, 1969 (in German).

607. Gould, R. W. F. and Raymer, W. G., Measurements Over a Wide Range of Reynolds Numbers of the Wind Forces on Models of Lattice Frameworks, National Physical Laboratory Sc. Rep. No. 5-72, Teddington, U.K., May 1972.

608. Ower, E., Wind Resistance of Lattice Girder Bridges, The Institute of Civil Engineers and the Institute of Structural Engineers, London, England, (1948).

609. Katzmayr, A. and Seitz, H., "Winddruck and Fachwerktuerme von quadratischen Querschnitt," Der Bauingenieur, Vol. 15, 1939, pp. 218-221.

610. Moll, R. and Thiele, F., "Windkanalversuche am Modell einer Stahlskelett Hochregalls zur Bestimmung des Wiederstandsbeiwertes nach DIN 1055 fur ein filligranartiges Bauwerk," Der Stahlbau, No. 3, March 1972.

611. Melling, R. J., Loads in Open Lattice Structures in Nominally Smooth Flow," Proceedings, 3rd Colloquium on Industrial Aerodynamics, Aachen, June 1978, C. Kramer and H.-J. Gerhardt, eds., Department of Aeronautics, Fachhochschule Aachen, Aachen, West Germany, 1978, pp. 107-127.

612. Sykes, D. M., "Wind Loads on Lattice Structures in Turbulent Airflow," Proceedings, 3rd Colloquium on Industrial Aerodynamics, Aachen, June 1978, C. Kramer and H. -J. Gerhardt, eds., Department of Aeronautics, Fachhochschule Aachen, Aachen, West Germany, 1978, pp. 129-148.

613. Jacobs, B. E. A., "Determination of Shielding Factors for Multiple Frame Structures," Proceedings, 3rd Colloquium on Industrial Aerodynamics, Aachen, June 1978, C. Kramer and H.-J. Gerhardt, eds., Department of Aeronautics, Fachhochschule Aachen, Aachen, West Germany, 1978, pp. 149-164.

614. Clow, D. G., "Loads on Open Lattice Structures: A Comparative Study," Proceedings, 3rd Colloquium on Industrial Aerodynamics, Aachen, June 1978, C. Kramer and H. -J. Gerhardt, eds., Department of Aeronautics, Fachhochschule Aachen, Aachen, West Germany, 1978, pp. 165-177.

615. Pagon, W. W., "Wind Forces on Structures: Plate Girders and Trusses," Journal of the Structural Division, ASCE, Vol. 84 No. ST4, July 1958, pp. 1711/1-1171/29.

616. Cohen, E. and Perrin, H., "Design of Multi-level Guyed Towers: Wind Loading, Journal of the Structural Division, ASCE, September 1957, pp. 1355/1-1355/29.

617. Cohen, E., "Wind Loads on Towers," Meterological Monographs, Boston, American Meterological Society, Vol. 4, No. 22, 1960, pp. 25-42.

618. Cohen, E., Vellozzi, J., and Suh, S. S., "Calculation of Wind Forces and Pressures on Antennas," Annals of the New York Academy of Sciences, Vol. 116, Art. 1, New York, N.Y., June 26, 1964, pp. 161-221.

619. Guide for Design of Steel Transmission Towers, prepared by the Task Committee on Tower Design of the Committee on Analysis and Design of Structures of the Structural Division, American Society of Civil Engineers, ASCE Manual and Report in Engineering Practice, New York, N.Y., 1971.

620. Structural Standards for Steel Antenna Towers and Antenna Structures, RS-222-C, Electronic Industries Association, Engineering Department, Washington, D.C., March 1976.

621. Design Manual - Structural Engineering, NAVFAC DM-2, Department of the Navy, Naval Facilities Engineering Command, Alexandria, VA, October 1970.

622. Antenna Towers and Antenna Supporting Structures, CSA Standard 537-1976, Canadian Standards Association, 1976.

623. Code of Basic Data for the Design of Buildings, Chapter V. Loading Part 2. Wind Loads, CP3, Gr 10, British Standards Institution, London, U.K., 1972.

624. Wilhoite, G. M., "Wind Recommendations for Transmission Towers," Journal of the Structural Division, ASCE, Vol. 98, No. ST2, February 1972, pp. 447-463.

7. GUYED TOWERS

In the past several decades, the needs of the broadcast industry, telecommunications and national defense have resulted in a substantial increase in the construction of multilevel guyed towers of great height. By supporting the shaft at suitable intervals with high strength structural strand guys, a highly economical structure results when land is available to provide adequate anchors for the guys.

In general, guyed towers have trussed structural steel shafts of triangular cross section. The shapes of the individual members are usually selected on the basis of their availability and adaptability to the manufacturing process of the fabricator. Members of high strength material and of circular cross section are generally used in order to reduce wind loads.

To reduce sag and increase stiffness, the guys usually receive a substantial prestress during erection. The points of attachment of the guy at the shaft and the guy anchorages are also located to optimize the shaft design by producing, as nearly as possible, a uniform displacement of the shaft under wind loading, so that a straight line can be passed between the deflected guy attachment points.

Apart from fixed loads due to the dead weight of the tower and the initial tensions in the guys, the primary loads on guyed towers are due to wind loads and ice loads. The loads due to wind pressure are greatly affected by the ratio of the solid area to the total enclosed area of a face of the shaft (i.e., the solidity ratio) and by the cross sectional shape of the individual members of the tower. In addition to wind forces on antennas, reflectors and other equipment mounted on the tower, electrical cables, co-ax lines and wave guides often add significantly to the overall wind load. To the extent that ice accretion increases both the dead load and the tributary wind area, ice loading is doubly important. Although many failures of guyed towers have been attributed directly to ice loading, little quantitative information exists regarding probable ice thickness for a given locality.

The governing standard for the design of towers for the broadcast industry in the U.S.A. is The Electronic Industries Association Standard EIA-RS-222C (Ref. 701). A discussion of this standard relative to wind loads for the design of towers is presented in Chapter 6 of this report. Wind forces on the guys and dynamic effects of wind on the guys are discussed in this Chapter. Also addressed in the material that follows is the dynamic response of tall guyed towers to turbulent wind forces. Procedures and data for performing a dynamic analysis of tall guyed towers to wind forces may be found in References 701 through 711, and in Chapter 4 of this report. Basic structural analysis of guyed towers using modern methods is covered in References 712 through 718.

Tower Guys

The drag and lift coefficients for guys (702) are shown in Figure 1 as a function of angle of attack of the wind. The angle of attack is the angle in space between the wind and the guy. The lift acts in a direction perpendicular to the wind and in the plane containing the wind vector and the guy. The drag acts in the direction of the wind.

The data in Figure 1 is based on tests on long, slender members of round shape. Note that, at an angle of 45 degrees, for example, the drag and lift coefficients are not equal. The drag coefficient is slightly higher, which implies a frictional drag along the length of the guy. This frictional drag theoretically exists at other angles but is small and may be due to experimental error. In any event, the influence of frictional forces would not appear to be significant.

Dynamic Effects - Guys

Wind vibration of tower guys is of two types: (1) aeolian or singing and (2) galloping or dancing. Aeolian vibrations are induced by alternating vortices shed in the wake of the guy. The motion is characterized by high frequency oscillations or singing. Aeolian vibrations can be controlled by the use of mass tuned vibration dampers (for Stockbridge type, see Figure 2) which absorb and dissipate high frequency guy vibration. In general, the higher the guy initial tension, the more prone they are to aeolian vibration. Conversely, the lower the tension, the more prone the tower is to overall lateral buckling or instability. In general, initial tensions in the range from about 8 percent to 15 percent of the guy breaking load are satisfactory, although Stockbridge type dampers should always be provided, particularly for tall towers in excess of several hundred feet in height.

The other type of guy vibration, known as galloping or dancing, consists of a low frequency, high amplitude vibration. Galloping of guys is usually associated with a coating of ice on the guys which changes the aerodynamic shape and renders "self-excited" vibrations possible. However, ice loading is not a necessary condition for the occurrence of galloping, as guys on tall guyed towers are often observed to gallop severely under conditions of gusty winds and no ice.

Galloping can be minimized by careful installation of a snubber and damper system situated near the lower end of the guy. One such system (Figure 3) consists of a tie-down attached to the lower portion of the guy by a pulley. The tie-down, in turn, contains a standard, automotive type shock absorber. Stockbridge dampers are ineffective in controlling galloping.

Guy vibration leads to fatigue failures in insulators and connections as well as in the individual wires comprising the guy

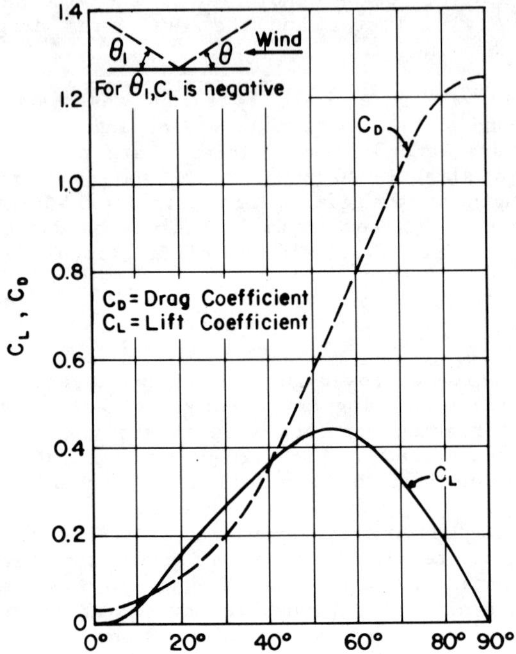

Figure 1 Drag and Lift Coefficients for Guys

Figure 2 Stockbridge Dampers

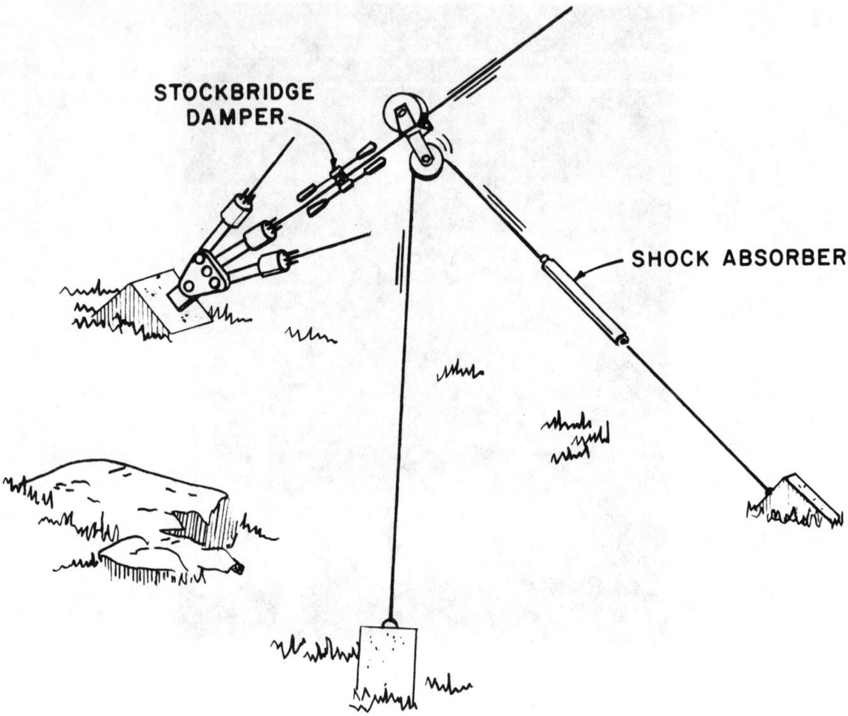

Figure 3 Guy Snubber

strand. If the vibrations go unnoticed, or uncontrolled, they eventually lead to premature failure of the entire tower.

Dynamic Effects - Waveguide

Individual members of trussed shafts of guyed towers, including co-ax lines, power lines, and so forth, are usually so stiff that vibration of these individual members by vortex shedding or galloping is generally not a problem and does not warrant investigation. Large waveguides on tall towers, however, are generally spring supported -for thermal expansion - and the flexibility introduced by the supports renders the waveguide vulnerable to wind induced oscillations. Large waveguides are generally circular or rectangular and may be mounted either inside the tower or external to a face. External mounting of large waveguides often eliminates interface problems with ladders and framing members on the inside of the tower, and simplifies erection and tuning of the waveguide. However, external hanging of large waveguides introduces undesirable torsional loadings which must be considered in design.

Normally, large waveguides are susceptible to Karman vortex induced oscillations at relatively low wind speeds, less than about 25 mph, corresponding to Strouhal numbers in the range from 0.1 to 0.2, depending upon the cross-sectional shape of the waveguide. Normally, it is impractical, and infeasible, to suppress such oscillations by use of spoilers. Practical solutions to the vortex shedding problem are achieved by use of frequently spaced lateral supports or guides for a waveguide (such supports must not impede vertical thermal expansion or contraction), or by use of vibration snubbers or dampers to suppress horizontal oscillations. In one instance, satisfactory results were achieved by tying-off the waveguide laterally with nylon rope.

Tall guyed radio and television towers usually have three-way guying at each guy level (no torque arms) and are torsionally very flexible. In such instances, eccentric mounting of a large waveguide can induce dangerous torsional oscillations of the tower itself. On a certain 1500-foot tower with an eccentric rectangular waveguide over the full height of the tower, torsional oscillations of the tower occurred repeatedly at wind-speeds above 30 mph for a particular wind direction. The oscillations were diagnosed as "torsional galloping" wherein the change in angle of attack of the wind relative to the waveguide (due to the torsional twist of the tower shaft) increased the dynamic lift coefficient for the waveguide, and hence the lateral force and torque in the system. The oscillations proved to be dangerous to the safety of the tower, necessitating a costly relocation of the waveguide to the centroid of the tower shaft.

Buffeting by Gusts

One of the most difficult problems in the design of guyed towers is the assessment of the effect of turbulent wind forces on the design quantities of interest, including bending moments and shears in the mast, as well as maximum guy tensions and axial loads. The current practice of treating gust loading on such structures is to multiply the "design" wind speed or pressure by a gust factor and to use the resulting load in static analysis. This approach is generally considered valid for buildings and other freestanding structures which tend to respond to gust loads in principally the fundamental mode of vibration. However, in the case of guyed towers, the higher modes of vibration make a significant contribution to the total dynamic response, thereby invalidating the simple gust factor approach for determining equivalent static wind forces.

One approach to the problem which was suggested by Cohen (702) is to use pattern wind loads in the spans between guy levels, similar to the procedure used in varying live loads on floor systems for calculating envelopes of maximum and minimum bending moments and shears. While this approach is considered rational and lends itself to the current methods of static analysis widely used in the design of guyed towers, virtually no guidance is available for determining pattern loadings which are consistent with the turbulence characteristics of strong winds and the dynamic response properties of guyed towers.

The stochastic response of guyed towers to turbulent wind forces has been studied by Hartmann and Davenport (703), and Vellozzi (711). The analytical models used by each of the above investigators are essentially similar, except that the effect of wind loading on the guys was neglected in (703) and (704). This effect does not result in major errors for towers of low height - less than several hundred feet - but the error in response predictions for very tall towers is over several hundred percent.

Because of the presence of the guy cables, a guyed tower is geometrically non-linear. This poses major analytical problems in the determination of the system response to fluctuating loads. In analyses conducted to date, the mean response of the tower to the steady wind loading component is computed with standard, non-linear guyed tower computer programs. For determining the additional dynamic response to wind gusts, the tower is assumed to oscillate linearly about the static equilibrium position produced by the mean wind loading component. This assumption permits standard matrix mode-superposition techniques to be used in a stochastic dynamic analysis. A set of significant lowest system frequencies and associated vibration modes are then obtained by solving a standard eigenvalue problem in which the stiffness matrix is based on the stiffness of the guys at the mean tower deflection.

Various techniques are used in the mathematical modeling of the guys to determine the natural modes and frequencies. Hartmann and Davenport, (703) as well as Beiten (704) and Vellozzi (711) used a lumped parameter model of the guys at each level in order to account for the first mode of guy vibration. This model was originally proposed in (703). Other methods use linearized deflection theory equations for the guys as well as finite element techniques in order to evaluate the influence of higher guy vibration modes on the overall system natural frequencies. The extent to which the higher modes of guy vibration affect the bending response of guyed towers to wind turbulence has not been evaluated, although the influence of the higher guy modes in this regard would not appear to be significant. However, out-of-phase vibrations between the guys on the windward and leeward sides of tall towers result in significant bending moments in the tower shaft, particularly for the top portion of the tower. Inasmuch as out-of-phase vibrations of the guys at a given level cannot be taken into account in a single lumped parameter model for <u>all</u> the guys at that level, it becomes mandatory in a buffeting analysis that a distinct model of each guy be included. Only by inclusion of a distinct degree of freedom for each guy can the coherence between the wind loads on the individual guys be taken into account.

The minimum number of normal modes required to achieve satisfactory results depends upon the number of guy levels. For an n level tower with 3 guys at each level, a minimum of 3n normal modes must be included to account just for the fundamental mode of each guy. To account for the shaft modes, an additional n modes is sufficient, bringing the total number of modes to 4n. It has been found that the higher system modes of vibration make a significant contribution to the bending moments and shearing forces in the mast. In the case of mast deflection and guy tension, the influence of the higher system modes is less significant.

Based on observations of the actual behavior of guyed towers under storm conditions, including observations reported in (719), procedures currently used to predict the dynamic response of tall guyed towers to turbulent wind loads tend to overestimate the dynamic response. The reason for this conservatism, it appears, is that the geometric stiffness of the guys is non-linear and a change in the frequency of oscillations over the half period hinders the growth of vibrations.

REFERENCES

701. Structural Standards for Steel Transmitting Antennas, Supporting Steel Towers, EIA Standard RS-222C, Electronic Industries Association, November 1966.

702. Cohen, E. "Wind Load on Towers", Meteorological Monographs, American Meterological Society, Volume 4, No. 22, May 1960.

703. Hartmann, A. J. and Davenport, A. G., Comparison of the Predicted and Measured Dynamic Response to Structures to Wind, Eng. Sc. Res. Report ST-4-66, University of Western Ontario, London, Ontario, 1966.

704. Beitin, K. I., Dynamic Response of Guyed Towers to Wind Loading, ASCE Environmental Engineering Meeting, Chicago, Illinois, October 1969, Preprint 1032.

705. Davenport, A. G. and Steels, G. N.,"Dynamic Behavior of Massive Guy Cables", ASCE Journal, Volume 91, ST2, April 1965.

706. Thompson, W. T. and Barton, M. V. The Response of Mechanical Systems to Random Excitation, Journal of Applied Mechanics, Volume 24, No. 2, 1957.

707. Vellozzi, J. W and Cohen, E., Dynamic Response of Tall Flexible Structures to Wind Loading, Building Science Series 30, National Bureau of Standards (U.S.), November 1970.

708. Harris, R. I., Measurement of Wind Structures at Heights Up to 598 Feet Above Ground Level, Electr. Res. Assoc. Report No. SPI/T31, 1968. (Unpublished)

709. Davenport, A. G., The Buffeting of Large Superficial Structures by Atmospheric Turbulence, Annals of the New York Academy of Sciences, Volume 116, Art. 1, 1964.

710. Davenport, A. G, The Application of Statistical Concepts to the Wind Loading of Structures, Proc. Inst. Civ. Eng., 19, pp. 449-472, 1961.

711. Vellozzi, J. W., "Tall Guyed Tower Response to Wind Loading", Proc., 4th International Conference on Wind Effects on Buildings and Structures, Heathrow, 1975, Cambridge University Press.

712. Williamson, R. A., "Stability Study of Guyed Tower Under Ice Loads", Journal of the Structural Division, Volume 99, pp. 2391-2408.

713. Goldberg, J.E. and Caunt, J.T., "Stability of Guyed Towers", Journal of the Structural Division, ASCE, Volume 99, No. ST4, Proc. Paper 9683, April 1973, pp. 741-756.

714. Hull, F. H., "Stability Analysis of Multi-Level Guyed Towers", Journal of the Structural Division, ASCE, Volume 88, No. ST2, Proc. Paper 3091, April 1962, pp. 61-80.

715. Odley, E. G., "Analysis of High Guyed Towers", Journal of the Structural Division, ASCE,Volume 92, No. ST1, Proc. Paper 4671, February 1966, pp. 267-325.

716. Williamson, R. A., and M. N. Margoliin, "Shear Effects in Design of Guyed Towers", Journal of the Structural Division, ASCE,Volume 92, ST5, Proc. Paper 4951, October 1966, pp. 213-260.

717. Williamson, R. A., Stevenson, H. F. and Vander Pas, P. W., "Computer Design of a Tower Cable System", Paper 5017, Journal of the Structural Division, ASCE, Volume 92, No. ST6, December 1966, pp. 341-360.

718. Felix Rosenthal and Richard A.Skop, "Method for Analysis of Guyed Towers", Journal of the Structural Division, Volume 108, No. ST3, March 1982, pp. 543-557.

719. Savitskii, G. A., Calculations for Antenna Installations, Svyaz' Publishers, Moscow, 1978. (Published for the National Science Foundation, Washington, D.C., by Amerind Publishing Co. Pvt. Ltd., New Delhi, 1982; available from National Technical Information Service, Springfield, Virginia 22161).

8. WIND LOADS ON COOLING TOWERS

8.1 Introduction

The critical loading for the reinforced concrete cooling tower is wind pressure. A great deal of research has been stimulated by the Ferrybridge cooling tower failures of 1965 (see Ref. 801 and section 805 herein). This Chapter reviews the state of the art in the description of wind loadings on cooling tower shells and comments on the extent to which various design procedures reflect current knowledge in this area.

The descriptions of the mean and of the fluctuating component of wind pressure are discussed in sections 8.2 and 8.3, respectively. Design procedures specified in Refs. 802 and 803 are commented upon in section 8.4. Section 8.4 contains information on grouped towers.

Studies of the buckling of cooling towers are reported in Refs. 804, 805, and 806. Observations on tornado wind effects on a cooling tower are reported in Ref. 807.

8.2 Mean Wind Pressures

The circumferential and vertical distributions of mean pressures are influenced by the characteristics of the oncoming flow (variation of wind speed with height, scale and intensity of turbulence), the tower geometry and surface, and the Reynolds number (Re). Insight into the flow around cooling towers has been gained by the study of the characteristics of the flow around long circular cylinders (see Chapter 2).

A schematic diagram of the typical mean pressure differences across the shell around a tower is shown in figure 8.2.1 (The pressure difference across the shell is the algebraic sum of the external and internal pressure, and constitutes the effective loading of the tower shell.) Measurements indicate that, for Re corresponding to typical design conditions, e.g., Re ~ 10^7-10^8, hyperbolic cooling towers, like cylinders, have an external pressure rise centered on the incident meridian ($\theta = 0^o$) that extends over an approximately 80° arc. An external suction occurs over the remainder of the surface.

Wind tunnel measurements can provide incorrect information on the mean external pressure coefficient $C_p(z, \theta)$, defined as

$$C_p(z, \theta) = (p(z, \theta) - p_{st})/q(z)$$

in which

$q(z) = 1/2\, \rho u^2(z)$
ρ = density of air
u = mean velocity at elevation z
$p(z,\theta)$ = mean pressure at elevation z and angle θ
p_{st} = static pressure of undisturbed flow

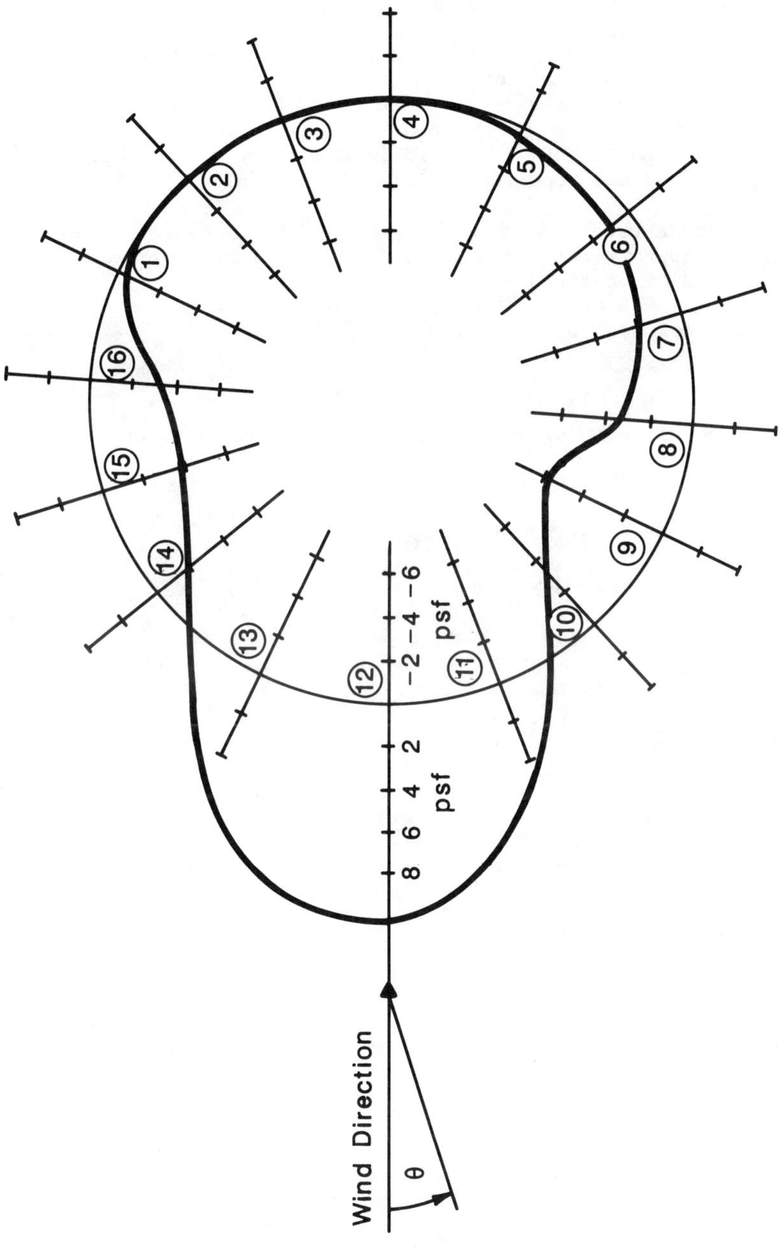

Figure 8.2.1 Sample Record of Mean Record of Mean Wind Pressure Differences Across Shell at Throat Level [808].

see figure 8.2.2 [809]. To obtain more reliable information, full-scale measurements have been performed. Measured values of C_p at the tower throat for several full-scale tests are shown in figure 8.2.3. The relatively large differences between the values for the tower of Ref. 810 and the other sets of values are explained by the absence of ribs on the external surface of that tower.

It was shown in Ref. 815 that the maximum suction and rear values of the C_p coefficients are functions of the elevation z. The maximum suction increases as z decreases (figure 8.2.4). Full-scale measurements of $C_p(z,\theta)$ nondimensionalized with respect to q(H) are presented in Ref. 816. Additional studies on the distribution of mean pressures are reported in Refs. 817 and 818.

A procedure based on Refs. 815, 817, 818 is now presented for determining approximate values of $C_p(z,\theta)$. The pressure rise coefficient ΔC_p is defined as:

$$\Delta C_p = C_{pb} - \min C_p \qquad (8.2.2)$$

in which

$\min C_p$ = value of C_p corresponding to maximum suction

$C_{pb} = -0.39\ (z/H)^{-2\alpha}$ = base pressure coefficient

H = total height of the tower

α = power law exponent (see Table 1.2.3).

It was shown in Ref. 815, from test results, that ΔC_p is independent of height, but it is influenced by the tower surface roughness, as shown in figure 8.2.5.

To obtain $C_p(z,\theta)$:

(1) Calculate k/a; use figure 8.2.5 to determine ΔC_p
(2) Calculate the maximum suction by $\min C_p = -\Delta C_p - 0.39\ (z/H)^{-2\alpha}$ (8.2.3)
(3) Use the following equations to determine the C_p distribution at each height z:

$$B = 1 - \min C_p \qquad (8.2.4a)$$
$$C = \ln(1/B)/\ln(\sin D\theta_0) \qquad (8.2.4b)$$
$$D = 90/\theta_1 \qquad (8.2.4c)$$
$$C_p(z,0) = 1 \qquad \theta = 0 \qquad (8.2.4d)$$
$$C_p(z,\theta) = 1 - B \sin^C D\theta \qquad \theta \le \theta_b \qquad (8.2.4e)$$
$$C_p(z,\theta) = C_p(z,\theta_b) \qquad \theta > \theta_b \qquad (8.2.4f)$$

Values of the angles θ_o, θ_1, and θ_b (defined in figure 8.2.6) are given for various terrain exposures and tower surface roughnesses in Ref. 839 and are reproduced from Ref. 111 in figure 8.2.7.

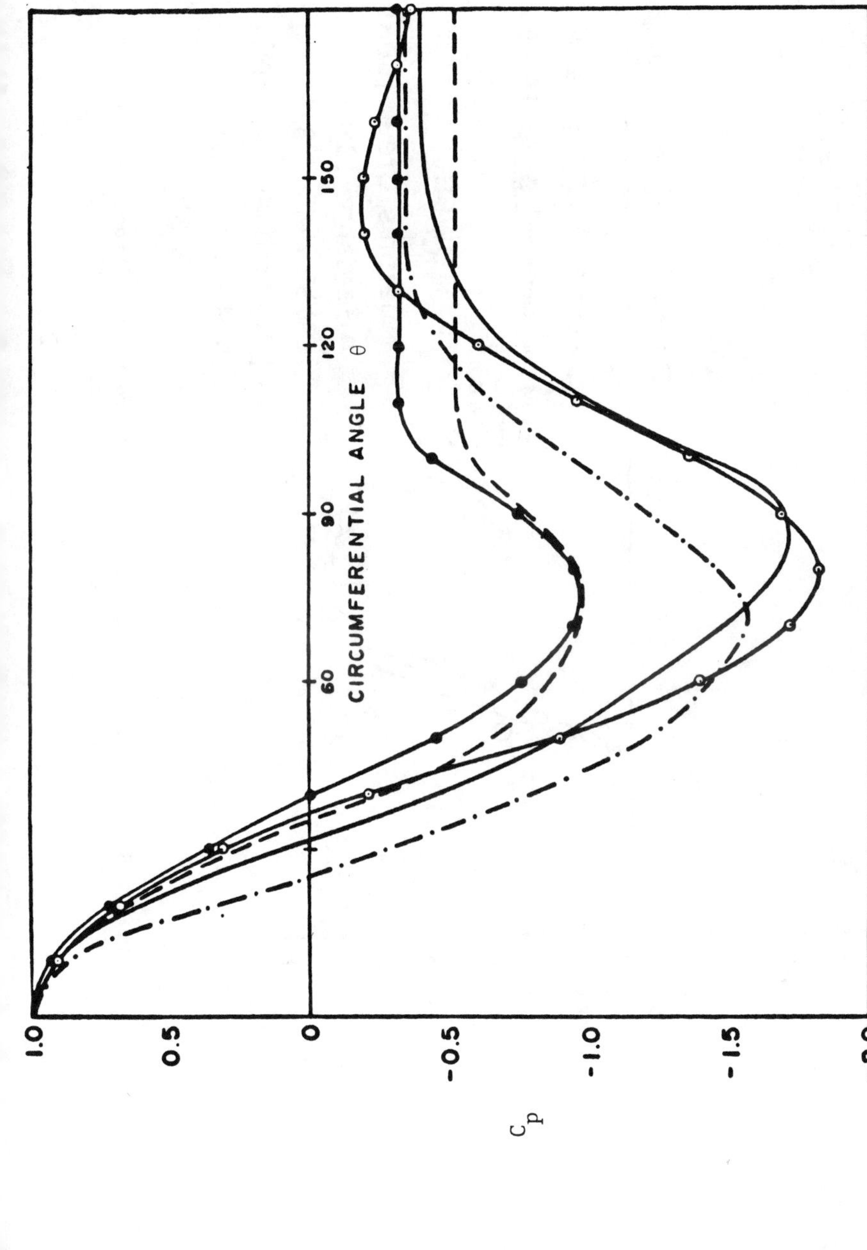

Figure 8.2.2 Mean Pressure Coefficient Around the Throat Section of a Hyperbolic Cooling Tower Obtained Tests [809].

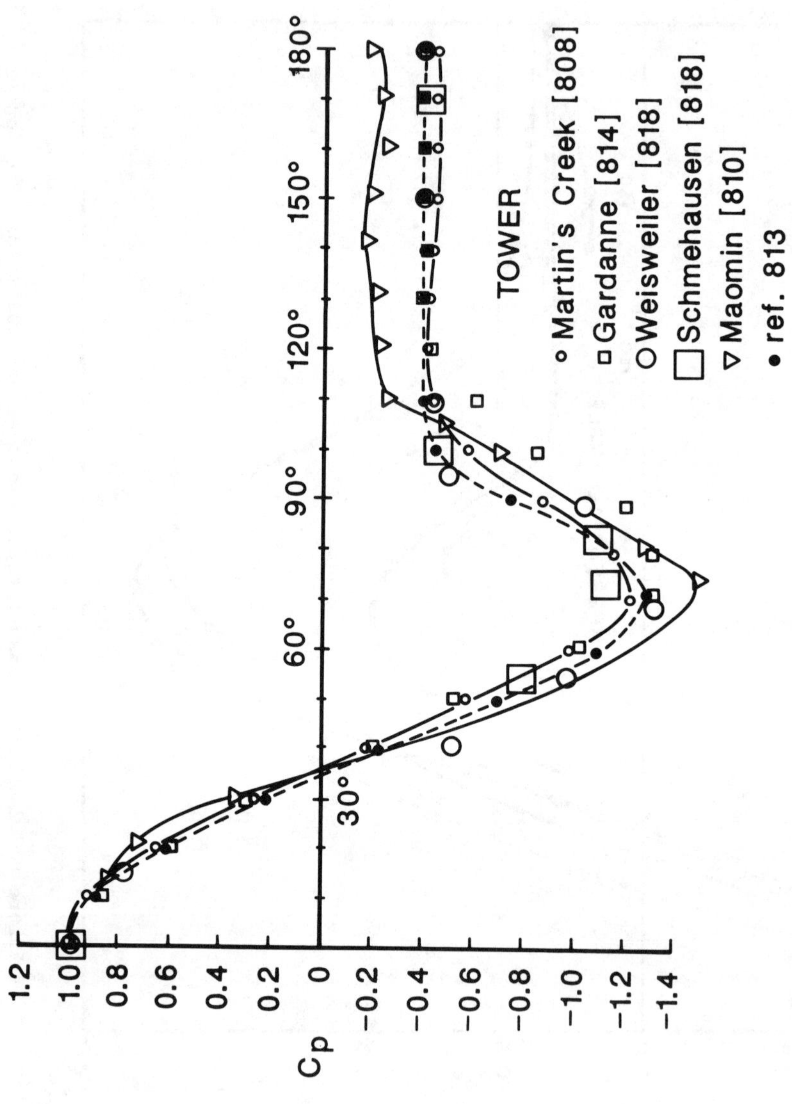

Figure 8.2.3 Mean Pressure Coefficient Around the Throat Section of a Hyperbolic Cooling Tower for Six Full-Scale Data Sets.

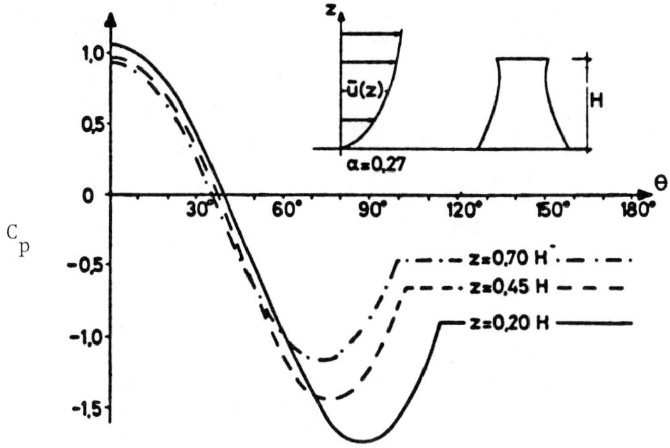

Figure 8.2.4 Height Dependence of C_p. From H. Proepper and J. Welsch, "Wind Pressures on Cooling Tower Shells", Proceedings, Fifth International Conference on Wind Engineering, Ft. Collins, Colorado, July 1979, Vol. 1, Pergamon Press, Oxford, 1980.

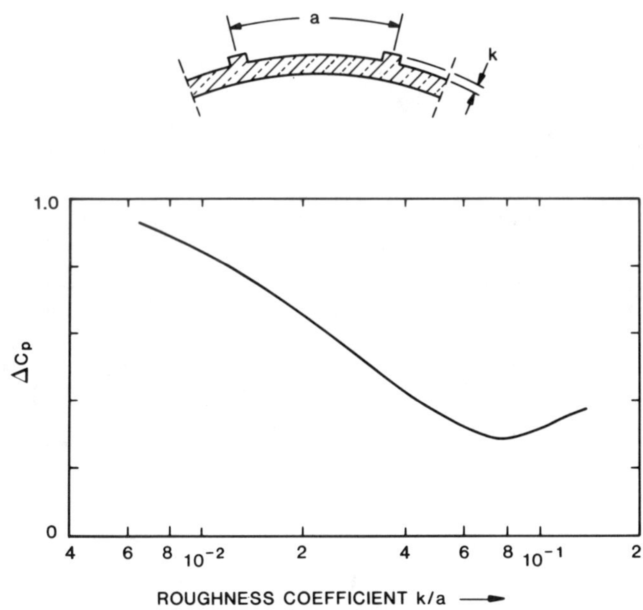

Figure 8.2.5 Influence of Surface Roughness on Pressure Rise Coefficients at Transcritical Reynolds Number [818].

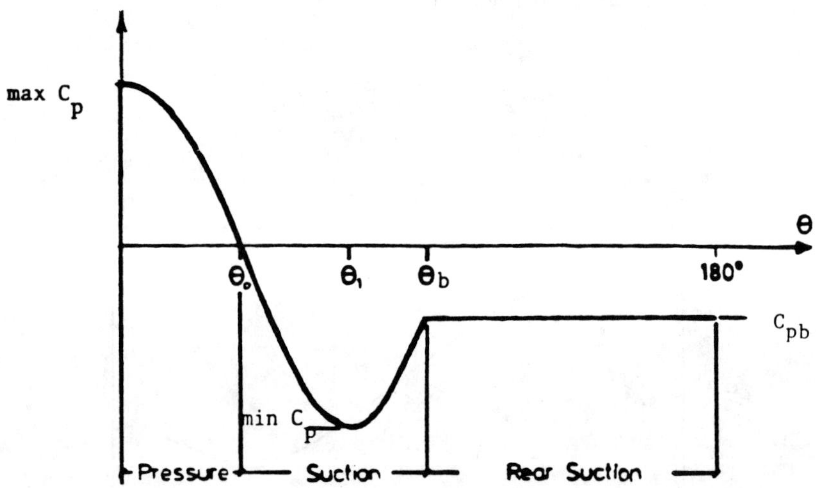

Figure 8.2.6 Angles θ_o, θ_1, θ_b.

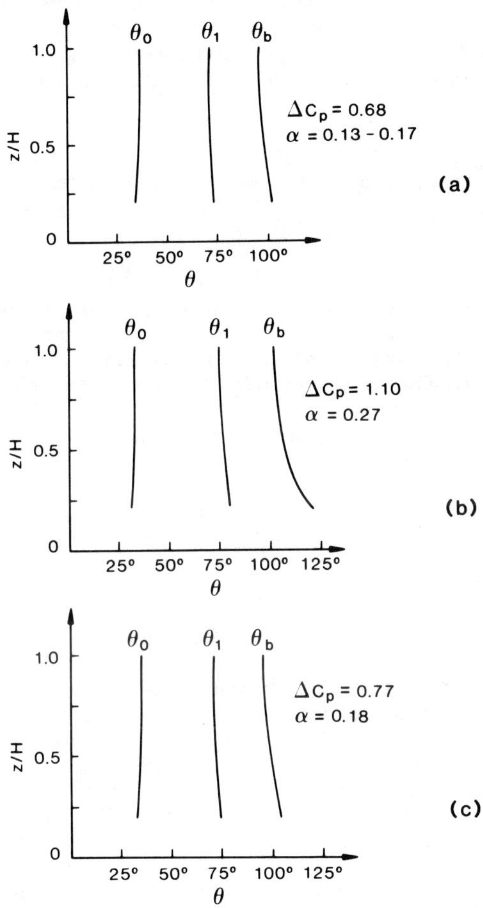

Figure 8.2.7 Angles θ_o, θ_1, θ_b

8.3 Description of Fluctuating Pressures

This section includes a review of information on the rms and the spectral and cross-spectral descriptions of the pressure fluctuations on cooling towers. Such information is necessary because the dynamic stresses in cooling towers are comparable in magnitude to the static values [808, 819, 824]. In addition, a brief review of a procedure for characterizing fluctuating pressures that employs time domain techniques is presented.

RMS of Pressure Fluctuations

The nondimensional fluctuating pressure coefficient C_p' is defined as

$$C_p' = \sigma_p/q(z) \qquad (8.3.1)$$

in which σ_p = rms pressure. Figure 8.3.1 shows three C_p' distributions obtained at the throat of the tower for model and full-scale tests. The model results are significantly different from the full-scale data. A peak in the rms pressures at approximately 90° is evident.

Both full-scale [814] and wind tunnel [809, 815] measurements have shown that the following linear relationship between C_p' and the intensity of turbulence, $I_u(z) = \sigma_u(z)/u(z)$, where $\sigma_u(z)$ = rms of longitudinal velocity fluctuation, holds at the stagnation point of the tower:

$$C_p' = \eta\ I_u(z) \qquad (8.3.2)$$

where $1.3 \le \eta \le 2.0$ according to Refs. 809, 814 and 815.

Subsequent full-scale measurements reported in Ref. 827 yielded results that cannot be modeled exclusively by a simple expression such as Eq. 8.3.2 for speeds $u(H) < 12$ m/s. For $u(H) > 12$ m/s, however, the results suggest that Eq. 8.3.2 can also be applied outside the stagnation point region, that is, Eq. 8.3.2 holds with $\eta = 1.5$ in the region $75^o < \theta < 95^o$; $\eta = 0.85$ in the region $95^o < \theta < 125^o$; and $\eta = 0.5$ in the region $125^o < \theta < 180^o$.

The influence of surface roughness coefficient, k/D, where D = diameter of the tower, on the circumferential distribution of C_p' has been noted in Refs. 809, 815 and 828. At an elevation near the throat, the C_p' values are lower for the rougher than for the smoother towers in the region $60^o < \theta < 120^o$.

A method of obtaining rms pressures at the stagnation point was suggested by the results obtained in Refs. 827 and 829. Let

$$\sigma_p^* = \sigma_p/\ \rho u\ \sigma_u \qquad (8.3.3)$$

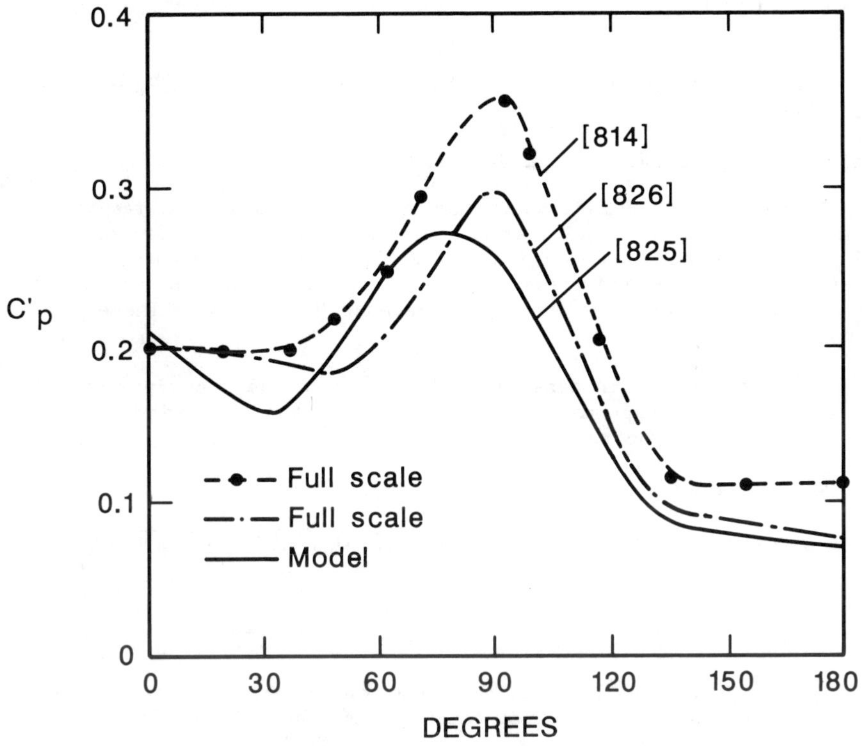

Figure 8.3.1 Fluctuating Pressure Coefficient around the Throat of a Hyperbolic Cooling Tower.

The following relation was found to hold for a typical cooling tower in atmospheric flow (Sageau [827]:

$$\sigma_p^* = \left(\frac{L_u^x}{D}\right)^{1/2} \tag{8.3.4}$$

Typical values for the integral turbulence length L_u^x are suggested in Section 1.2.2.1.

Spectral and Cross-Spectral Description

In solving the equations of motion for the structural response, it is convenient to describe the pressure loading by use of spectra and cross-spectra (see Section 1.2.2.3). For the coordinates s (meridional) and θ (circumferential), the pressure cross-spectra are defined in terms of various coherence functions. These are discussed below.

The cross-spectral densities of the pressures are represented for two regions, windward and wake. On the basis of wind tunnel tests, the cross-spectra are defined in Ref. 823 as follows:

(1) Wind region ($\theta < 100^o$) (8.3.5)

$$S_p(s,\theta,s',\theta',n) = R_v(s,s',n)\ R_f(\theta,\theta',n)\ \sqrt{S_p(s,\theta,n)\ S_p(s',\theta',n)}$$

(2) Wake region ($\theta > 100^o$) (8.3.6)

$$S_p(s,\theta,s',\theta',n) = R_v(s,s',n)\ R_f(\theta,\theta',n)\ \sqrt{S_p(s,\theta,n)\ S_p(s',\theta',n)}$$

Cross-spectra between windward and wake pressures are assumed to be negligible. The vertical coherence may be written

$$R_v(s,s',n) = \exp(-\beta_1\eta_1) \tag{8.3.7}$$

in which $\eta_1 = n\Delta s/u(z_g)$, Δs = vertical separation, and $u(z_g)$ = mean velocity at gradient height z_g, and $\beta_1 \simeq 7$ [823].

The frontal coherence may be written as

$$R_f(\theta,\theta',n) = C_2(\theta,\theta')\ R_r(\Delta\theta,\ n) \tag{8.3.8}$$

where $\Delta\theta = |\theta - \theta'|$

Examples of cross-correlation coefficients $C_2(\theta,\theta')$ in the windward region proposed in Ref. 823 are plotted in figure 8.3.2. It can be seen that according to figure 8.3.2 no correlation of fluctuating pressures for (θ,θ') is apparent in the wake region. Additional information on this correlation coefficient is available in [815].

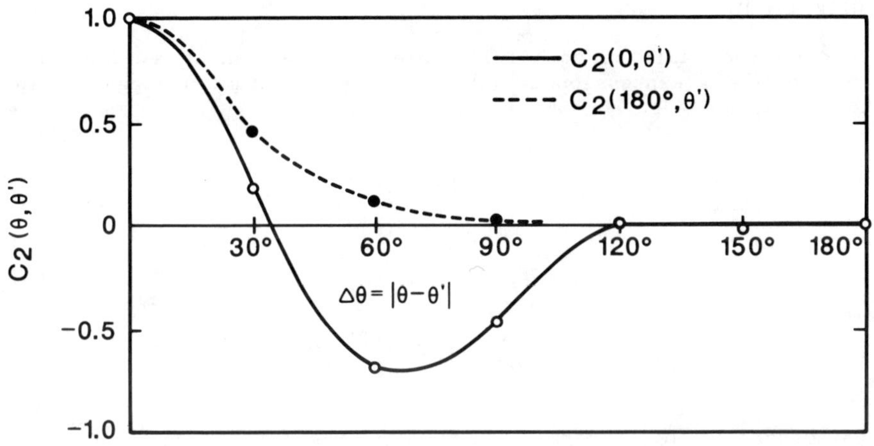

Figure 8.3.2 Cross-Correlation Coefficient at Throat Level [823].

The function $R_f(\Delta\theta, n)$ is given by

$$R_f(\Delta\theta, n) = \exp(-\beta_3 \eta_2^2) \qquad (8.3.9)$$

in which $\beta_2 \simeq 25$, $\eta_2 = nr\Delta\theta/u(z_g)$, and r = radius of the tower [823].

The rear coherence is given by

$$R_r(\theta,\theta', n) = \exp(-\beta_3 \quad \eta_2) \qquad (8.3.10)$$

in which $\beta_3 \simeq 11$ [823]. Alternative and, in some instances, more conservative expressions for the coherence function are recommended in [828]. Reference 823 contains plots of $R_r(\theta,\theta', n)$ and $R_f(\theta,\theta', n)$.

On the basis of both full-scale and wind tunnel measurements the following expression for the pressure spectrum is suggested in Ref. 815:

$$\frac{n \; S_p(\theta, n)}{\sigma_p^2(\theta, z)} = \frac{\beta_p(\theta) \; X_p(\theta)}{[1 + \nu_p(\theta) X_p^2(\theta)]^{\delta(\theta)}} \qquad (8.3.11)$$

in which

$$X_p(\theta) = [b_o(\theta)^{1/\alpha} \bullet (D/{}^{x}Lu)^{2/3}]^{1/a_1(\theta)} \frac{nD}{u(z)} \qquad (8.3.12)$$

parameters a_1, b_o, β_p are taken from figures 8.3.3a-c and

$$p(\theta) = [\beta_p(\theta)/0.115]^{1/\delta(\theta)} \qquad (8.3.13)$$

$$\delta(\theta) = [1 - a_1(\theta)]/2. \qquad (8.3.14)$$

${}^{x}Lu$ is the integral scale of turbulence (section 1.2.2.1) and α is the power law exponent (table 1.2.3).

Time Domain Description

A procedure for representing fluctuating loads in the time domain is described in Ref. 808. Given a stagnation point pressure record at the throat the procedure allows the calculation of scaling factors that account for the dependence of the fluctuating loads upon position on the tower surface. The stagnation point pressure record at the throat can be simulated from spectral information, (see e.g., section 2.1.3.3) [832], or by using the ARIMA (Auto Regressive Integrated Moving Average) models developed for full-scale tower data on the basis of the Box-Jenkins approach [835-855].

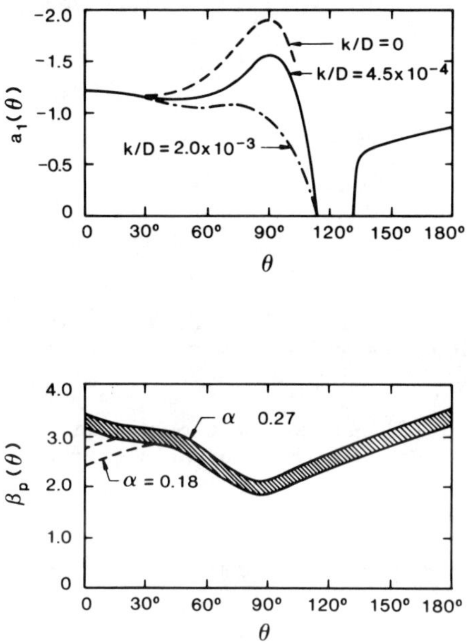

Figure 8.3.3 Parameters Used in Spectral Expressions for Pressure Fluctuations [815]. From H. Proepper and J. Welsch, "Wind Pressure on Cooling Towers and Shells", Proceedings, Fifth International Conference on Wind Engineering, Ft. Collins, Colorado, July 1979, Pergamon Press, Oxford, 1980.

8.4 Equivalent Static Loading

Equivalent static loadings, which account for the effects of both the mean and fluctuating pressures, can be estimated by following the simplified procedure presented in the report of the ACI-ASCE Committee 334 [802], or the procedure developed in Germany and described in Ref. 803.

ACI Procedure [802]

The equivalent static loading recommended by ACI Committee 334 is defined below:

$$q_z = K_z\ G\ q_{30}\ H_\theta \tag{8.4.1}$$

in which

q_z = the effective velocity pressure in psf at height z in feet above ground level along the windward meridian where the pressure is maximum

K_z = an exposure factor that establishes the vertical profile of wind speed and that depends upon the roughness of the terrain, taken from [8.36].

G = a gust response factor which depends upon the dynamic response of the structure to turbulence, which should be greater than unity.

H_θ = a coefficient for circumferential distribution of wind pressure obtained from wind tunnel tests or full-scale measurements.

and

$$q_{30} = 0.00256 u_{30}^2 \tag{8.4.2}$$

q_{30} = basic velocity pressure in psf at 30 ft above the ground level obtained from the basic wind speed (fastest-mile-of-wind) u_{30} in mph at 30ft above ground level with a 100-year return period [836].

Procedure Described in Ref. 818

According to Ref. 818, the wind pressures, $q(z,\theta)$, acting on a tower in open country may be defined as

$$q(z,\theta) = c(\theta)\ q_D(z) \tag{8.4.3}$$

in which

$c(\theta)$ = coefficient of circumferential distribution of pressures. For open country,

$$q_D(z) = 1.71 \left(\frac{z}{z_g}\right)^{0.23} \frac{1}{2} \rho\, u^2(z_g)\, \phi \qquad (8.4.4)$$

in which

z = elevation in meters,
z_g = gradient height assumed to be equal to 280 m,
ρ = 1.25 kg/m^3,
$u(z_g)$ = mean hourly wind speed at elevation z_g,
ϕ = resonance factor assumed to have the value $1 < \phi < 1.1$.

The distribution of the pressure coefficient, $c(\theta)$, plotted in figure 8.4.1 [817], depends on the surface roughness parameter. The parameter is defined as the ratio of the height of the ribs, k, to the distance between their centers, a.

8.5 Grouped Towers

Studies of grouped towers have been undertaken by Armitt [837] using wind tunnel and full-scale tests, and by Sageau [838] using full-scale measurements. The data of Ref. 837 show that locally both the mean and the fluctuating pressures can be affected significantly by the presence of neighboring towers. In certain instances, pressures can differ by as much as 35 percent from the corresponding pressures on isolated towers. Sageau [838] found similar results in his study of staggered towers. A lack of symmetry of the loads on the staggered tower was also noted. The effect of grouping was found to be significant in the case of the Ferrybridge tower failures, which were determined to have occurred at speeds of about 20 m/s as a result of tensions in the windward face of the towers [837].

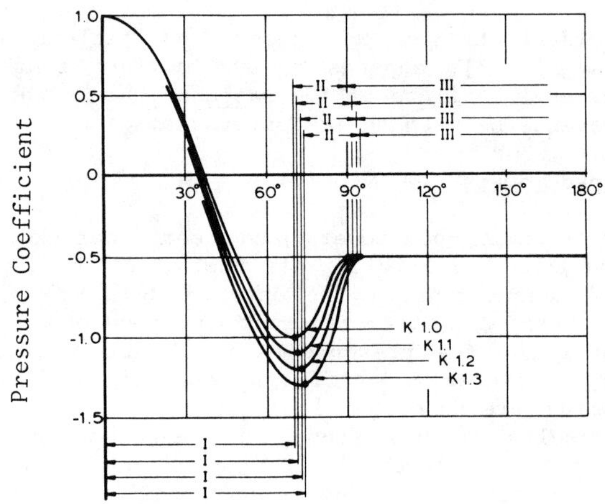

Figure 8.4.1 Distribution of the Pressure Coefficient c(θ) used in Ref. 817. K values denote surface roughness: K1.0, K1.1, K1.2, K1.3, corresponds to ratio k/a = 3×10^{-2}, and 8×10^{-2}, and 8×10^{-3}, respectively; I, II, and III denote the three pressure regions of the tower.

REFERENCES

801. Report of the Committee of Inquiry into Collapse of Cooling Towers at Ferrybridge, Monday, November 1, 1965, Central Electricity Generating Board, Her Majesty's Stationery Office, London, England, 1966.

802. "Reinforced Concrete Cooling Tower Shells: Practice and Commentary," ACI 334, IR-77, American Concrete Institute, Detroit, Michigan, 1977.

803. Zerna, W., Kratzig, W. B. and Mungan, I., "Cooling Tower Practice in Germany: State of the Art," ASCE Preprint 81-005, May 1981.

804. Abel, J.E. and Gould, P.L., "Buckling of Concrete Cooling Tower Shells," Concrete Shell Buckling, ACI Publication SP-67, American Concrete Institute: Detroit, 1981, pp. 135-231.

805. Billington, D.P. and Harris, H.G., "Test Methods for Concrete Shell Buckling," Concrete Shell Buckling, ACI Publication SP-67, American Concrete Institute: Detroit, 1981, pp. 187-231.

806. Sageau, J.F. and Robert, M., "Etude Experimentale des Modes de Rupture des Grandes Tours de Refrigération," Electricité de France,Direction des Etudes et Recherches, 6 quai Watier, Chatou, France, 1982.

807. Fujita, T.T. "Tornado Wind Effects on The Grand Gulf Cooling Towers," Proceedings of the 2nd Specialty conference on Dynamic Response of Structures: Experimentation, Observation, Prediction and Control, January 15-16, 1981, pp. 555-566.

808. Sollenberger, N.J., Scanlan, R.H., and Billington, D.P., (1980) "Wind Loading and Response of Cooling Towers," J. Structural Engineering, ASCE, 106(3), 601-622.

809. Hashish, M. G., "Wind Response of Hyperbolic Cooling Towers," Ph.D. Thesis, University of Western Ontario, London, Canada, September 1973.

810. Sun, T-.F. and Zhou, L-.M., "Wind Pressure Distribution on a Ribless Hyperbolic Cooling Tower," Proceedings of the Sixth Internationsl Conference on Wind Engineering, Session 11, Vol. 2, 126, Gold Coast, Australia, March 21-25, 1983.

811. "Wind Forces on Structures," Transactions, ASCE, Vol. 126, Part 2, 1961.

812. Walser, A. and Gurfinkel, G., "Structural Design of Hyperbolic cooling Towers," IASS Proceedings, Shell Structures and Climatic Influences, Calgary, Canada, 1972.

813. Niemann, H. J., "Zur Stationaren Windbelastung rotationsymmetrischer Bauwerke in Bereich transkritischer Reynoldzahlen," Report 71-2, Inst. fuer Konstruktiven Ingenieurbau Ruhr-Universitat Bochum, March 1971.

814. Sageau, J. F., "In-Situ Measurement of the Mean and Fluctuating Pressure Fields Around A 122 Meters, Smooth, Isolated Cooling Tower," Direction Des Etudes et Recherches, Electricite de France, 6 quai Watier, Chatou, France, September 11, 1979.

815. Proepper, H. and Welsch, J., "Wind Pressures on Cooling Tower Shells," Proceedings of the 5th International Conference on Wind Engineering. Ft. Collins, Colorado, USA, July 1979, Vol. 1, pp. 465-478.

816. Sageau, J. F. and Grovalet, Y., "Charges moyennes et contraintes induites par le vent sur la coque des grands refrigerants," Rejets de chaleur a l'atmosphere, Societte Hydrotechnique de France, XVI Journees de l'Hydraulique, Paris, 16-18, September 1980, pp. 4.1.8.

817. Harnach, R. and Niemann, H. J., "The Influence of Realistic Mean Mean Wind Loads on the Static Response and Design of High Cooling Towers," Eng. Struct., 1980, Vol. 2, January, pp. 21-34.

818. Niemann, H. J., "Wind Effects on Cooling Tower Shells," Journal of the Structural Division, ASCE, Vol. 106, No. ST3, Proc. paper 15251, March 1980, pp. 643-661.

819. Abu-Sitta, S. H. and Hashish, M. G., "Dynamic Wind Stresses in Hyperbolic Cooling Towers," Journal of the Structural Division, ASCE, Vol. 99, No. ST9, Proc. Paper 9989, September 1973, pp. 1823-1835.

820. Billington, D.P. and Abel, J., "Design of Cooling Towers for Wind," Proceedings of the Specialty Conference, "Methods of Structural Analysis," University of Wisconsin - Madison, August 22-25, 1976, pp. 242-267.

821. Steinmetz, R. L., Billington, D.P. and Abel, J., "Wind Loading and Response of Cooling Towers," Journal of the Structural Division, ASCE, Vol. 104, No. STI, Proc. Paper 134, Jan. 1978, pp. 35-53.

822. Isyumov, N., Abu-Sitta, S. H. and Davenport, A.G., "Approaches to the Design of Hyperbolic Cooling Towers Against the Dynamic Action of Wind and Earthquake," International Association of Shell Structures Bulletin, No. 48, March 1972, pp. 3-22.

823. Hashish, M. G. and Abu-Sitta, S. H., "Response of Hyperbolic Cooling Towers to Turbulent Wind," Journal of the Structural Division, ASCE, Vol. 100, No. ST5, Proc. Paper 10542, May 1974, pp. 1037-1051.

824. Basu, P. K. and Gould, P. L., "Cooling Towers Using Measured Wind Data," Journal of the Structural Division, ASCE, Vol. 106, No. ST3, Proc. Paper 15259, March 1980, pp. 579-600.

825. Davenport, A. G. and Isyumov, N., "The Dynamic and Static Action of Wind on Hyperbolic Cooling Towers," Vol. 1, Research Report No. BLWTI-66, University of Western Ontario, London, Canada, 1966.

826. Ruscheweyh, H., "Wind Loadings on Hyperbolic Natural Draught Cooling Towers," J. Ind. Aerodyn., 1 (1976), pp. 355-340.

827. Sageau, J. F., "Champ de pression fluctuant sur la coque de refrigérant 5 Ouest de la centrale du Bugey," Electricité de France, Direction des Etudes et Recherches, 6 quai Watier, Chatou, France, December 30, 1981.

828. Pirner, M., "Wind Pressure Fluctuations on A Cooling Tower," J. Wind Eng. and Ind. Aerodynamics, Vol. 10, No. 3, Dec. 1982, pp. 343-360.

829. Durbin, P. A. and Hunt, J.C.R., "On Surface Pressure Fluctuations Beneath Turbulent Flow Round Bluff Bodies," J. fluid Mechanics (1980), Vol. 100, Part 1, pp. 161-184.

830. Vickery, B. J. and Kao, K. H., "Drag or Along-Wind Response of Slender Structures," Journal of the Structural Division, ASCE, Vol. 98, No. ST1, January 1972.

831. Kawarabata, Y., Nakae, S., and Harada, M., "Some Aspects of the Wind Design of Cooling Towers," Proceedings of the Sixth International Conference on Wind Engineering, Session 11, Vol. 2, Gold Coast, Australia, march 21-25, 1983.

832. Shinozuka, M., "Applications of Digital Simulation of Gaussian Random Processes," Lecture Notes for the Course, "Random Excitations of Structures by Earthquakes and Atmospheric Turbulend," Presented at the International Center for Mechanical Science3s, Udine, Italy, July 12-16, 1976.

833. Box, G. P. and Jenkins, G. M., Time Series Analysis: Forecasting and Control, Holden-Day: San Francisco, California, 1976.

834. Fortier, L. J. and Scanlan, R. H., "A Cooling Tower Wind Loading Model Based on Full-Scale Data," Proceedings of the Sixth International Conference on Wind Engineering, Session 11, Vol. 2, Gold Coast, Australia, March 21-25, 1983.

835. Reed, D. A. and Scanlan, R. H., "Cooling Tower Wind Loading," Proceedings of the 4th U.S. National Conference on Wind Engineering Research, Seattle, Washington, July 26-29, 1981, Vol. 1, pp. 254-261.

836. "Building Code Requirements for Minimum Design Loads in Buildings and Other Structures," ANSI A58-1982, American National Standards Institute, New York, 1982.

837. Armitt, J., "Wind Loading on Cooling Towers," Journal of the Structural Division, ASCE, Vol. 99, No. ST9, Proc. Paper 15282, March, 1980, pp. 623-641.

838. Sageau, J. F., "Caractérisation des Champs de Pression Moyens et Fluctuants à la Surface des Grands Aérorefrigerants," Electricité de France, Direction des Etudes et Recherches, 6 quai Watier, Chatou, France, July 1979.

839. Welsch, J., "Der Einfluss des Windprofiles auf die statischen Windbeanspruchungen von rotationshyperbolischen Kuehlturmschalen," Lehrstuhl I, Institut fuer konstruktiven Ingenieurbau, Ruhr-Universitaet Bochum, Feb., 1978.

840. Niemann, H.J., "Cooling Tower Wind Loading and Response," U.S. -German Joint Seminar on Structural Reliability, Columbia University, N.Y., N.Y., May 20, 1983.

9. BRIDGES

The state-of-the-art is that, with great confidence, nearly all wind induced problems with bridges can be prevented in the design phase. This is not meant, however, to imply that there is no need for further research in this area. Failure can be prevented, but the analysis for certain modes of failure can be overly conservative. The advances yet to come are refinements in the experimental and analytical procedures that will reduce excessive conservatism.

All types of bridges are susceptible to wind-induced problems. Typically, wind-induced problems are most common and to be expected for suspension bridges, cable stayed bridges, and long-span box girder bridges, and particular attention should be given to all types of bridges under construction (903). During construction, bridges and/or bridge components, have more flexibility and have more degrees of freedom than they are designed to have in their completed states. Consequently, wind induced problems may occur for wind velocities that are truly likely to occur. Most likely, temporary measures may be used to prevent the occurrence of problems which might arise during the construction phase. Specifically cables, braces, and dampers can be added as temporary measures to prevent excessive wind induced motions during construction.

Certain generalizations can be made for most bridges and for most anticipated problems. First, wind-induced problems with bridges are typically most severe when winds are perpendicular to the axis of the bridge, or very nearly so. Unless stated otherwise, in the discussion which follows, it will be assumed that winds are horizontal and perpendicular to the axis of the bridge. Second, wind-induced problems typically are associated with the bridge deck and its structure. In this chapter the discussion will concern wind loads on, and the corresponding response of, bridge decks of suspension bridges, cable stayed bridges, and long span box girder bridges. Again, it is not meant to imply that other wind induced problems do not exist. They do exist less frequently and they may be treated using methods outlined in other sections of this report. However, the most common problems, and the specialized methods to solve those problems, are associated primarily with the decks of those bridge types mentioned. Furthermore, the types of wind-induced problems on bridges discussed in this chapter are those involving structural performance. Wind loads on trucks, trailers, automobiles, and pedestrians can be large and dangerous, and can present real problems. These will not be discussed in this chapter, however. Those problems which are addressed and which occur most frequently are the following:

- Torsional divergence - the monotonically increasing, quasi-static, angular distortion of the bridge deck about its centerline;

- Lock-on vortex induced vibrations - the forced vibration due to periodic shedding of vorticies at a natural frequency of the bridge;

- Flutter and galloping - the self-induced, divergent, oscillatory vibration of bridge deck which may lead to the collapse and total failure of the bridges; and

- Buffeting - the forced vibration of the bridge deck caused by the turbulence in the oncoming flow.

Bridge deck hangers, slender bridge towers, and other slender and flexible elements of bridges may also vibrate in the wind (904). These vibrations are not unique to bridges, and can be analyzed and prevented as discussed elsewhere in this report.

All four of the most common problems involve bluff body aerodynamics. Bluff body aerodynamics is extremely complex, very little of which can be explained theoretically from basic principles. The bulk of all analyses, therefore, relies upon results of experimental studies. Most often these experiments are specific studies for the project in question. All problems discussed in detail in this report consequently rely heavily upon results of experimental studies.

This chapter is divided into five parts. Because of the importance of experimental procedures, a separate part is devoted to the discussion of state-of-the-art experimental procedures. The remaining four parts are devoted to discussions of the state-of-the-art methods of analysis associated with the four most commonly expected wind induced problems: torsional divergence, lock-on vortex induced vibrations, flutter and galloping, and buffeting.

9.1 WIND TUNNEL TEST PROCEDURE

Experimental studies are most often performed in atmospheric boundary layer wind tunnels, traditional wind tunnels used for studies of aircraft aerodynamics, and wind tunnels designed specifically for the analysis of bridges. State-of-the-art wind tunnel test procedures can be divided by model type (full model, taut strip, section model) and by the condition of the flow (smooth or turbulent). Both divisions are discussed.

9.1.1 Full Bridge Models

Since the inception of wind tunnel testing of bridges, there have only been a handful of tests on models of full bridges (905, 906, 908). The advantages of full bridge model tests are: 1) they represent the proper interaction of bridge deck piers, towers, abutments, and cables (if they exist); 2) they can represent the proper flow distortions on all parts of the bridge if the surrounding topography is modeled as well as the bridge (it is worthy of note that the original, unstiffened configuration of the Golden Gate Bridge was most likely saved from more severe consequences because of the downward distorted flow on the north half of the bridge from the interference of the Marin headlands); and 3) in certain cases the model scale permits the proper turbulent structure of the wind to be modeled well. Disadvantages of full bridge wind tunnel tests are the following: 1) the wind tunnel models are very costly; 2) the model must be scaled geometrically and dynamically (see scaling considerations discussed elsewhere in this report); 3) the model is not readily modified if the original configuration proves to be unacceptable; and 4) the model scale must necessarily be small and, therefore, Reynolds numbers must also be small.

Model expenses are high because not only must the geometry of the full bridge be modeled but so too must all quantities that would affect the dynamic response, i.e., the mass, the stiffness, and the damping. Since all of these quantities do not scale linearly with length, great care must be taken to model all quantities preoperly. In most cases a backbone structure is provided with a lightweight geometrically modeled shell consisting of separated rigid segments.

To be restricted to small Reynolds numbers is also a problem because Reynolds number dependence of the results can be significant. Typically, aerodynamic forces on sharp-edge stationary models are not Reynolds number dependent (907). However, the time dependent forces on typical bridge decks (whether they have sharp or rounded edges) are highly dependent upon the turbulence in the wake behind the bridge deck, and that wake turbulence structure is Reynolds number dependent. An effort should be made to reduce the Reynolds number mismatch between model and prototype, but the discrepancy usually remains large under ordinary methods of testing. Furthermore, it has been shown that the time-dependent forces on some bridge decks are very sensitive to small changes in detail such as the addition of slots in the roadway, barriers and railings with varying porosities (910). Flows in and through these small openings can be highly Reynolds number dependent as well. In addition, these differences in geometric details (which have been shown to have significant effects on the aerodynamic forces) are often smaller than the gaps and geometric mismatches that are inevitable when a continuous, flexible structure is modeled approximately as a series of elastically connected rigid segments. Very small scale inevitably means loss of detail, and small changes in detail have been shown to cause significant variations in aeroelastic behavior.

These disadvantages plus the lack of flexibility in full bridge model testing made it not particularly desirable. Whenever full bridge model testing is used, it is common to use other testing methods as well.

9.1.2 Taut Strip Models

A slight simplification to full bridge model testing is taut strip model testing (906, 908). As the name implies the stiffness properties of the bridge are modeled with wires tightly strung across the wind tunnel. Over these tight wires are placed light deck sections which model the inertial and geometric configuration of the bridge deck. The taut strip model is, in essence, a model of the main span of the bridge.

Advantages of the taut strip model over the full bridge model are the following: 1) it is less expensive; and 2) its structural properties can be changed with greater ease than those for a full bridge model.

The taut strip model loses some precision in modeling flows at the towers, pier, and abutments as compared to a full bridge model. It can, however, be argued that loss of precision of airflow in those areas creates insignificant variations in the results. Taut strip models are smaller than what would be desirable; they are likely to be made of rigid segments elastically joined, and they inevitably will not have the precision in detail that is desired. Taut strip models are not commonly used.

9.1.3 Section Models

By far the most common method of testing uses section models of the bridge deck and the bridge deck structure. A large scale rigid model of a section of the bridge deck is made of a section of the bridge deck is made and supported by springs so that the lowest-mode dynamic character of the bridge deck section is modeled (910). The basic use of most section models is as a test device to adjust the cross-section of a bridge deck to an acceptably stable aerodynamic form. As will be discussed in the section on flutter, section models are not necessarily designed to be models of actual bridge behavior but often are merely "analog computers" that can be used to extract experimentally time-dependent aerodynamic forces which then are used in in analytical models of the full bridge. Consequently, masses and stiffnesses need not always be modeled exactly.

Not all section models are supported on springs. Some are driven, via a linkage arrangement, through a prescribed oscillation (910). Inertial forces of the model have then been separated from aerodynamic forces ingeniously by simultaneously driving a similar model outside of the airstream, only the differences in loads between section model and driving mechanism being recorded as the aerodynamic components.

Advantages of section model testing are the following: 1) the models themselves are inexpensive (the driving mechanism can be very expensive but the spring supported mechanism adds little to the overall expense); 2) the models are of large enough scale so that fine details can be modeled with accuracy; and 3) geometric and dynamic properties of the section model and its supporting system can be modified with ease.

There are few disadvantages of section model tests. There is some lack of precision modeling of the true aerodynamic loading near abutments, piers, and towers. However, at these locations the bridge deck is typically fixed, or distortions are small. Therefore, contributions to generalized aerodynamic forces and moments near the bridge supports can often be considered minimal, and errors introduced in the total generalized aerodynamic forces and moments from these contributions is insignificant.

State-of-the-art experimental testing is, in fact, section model testing for the most part. Results from full bridge model or taut strip model tests alone would be suspect because of the lack of fine detail and the low Reynolds number. For greatest confidence in results, full bridge model and taut strip model tests should be suplemented with section model tests.

9.1.4 Smooth vs. Turbulent Flow Tests

The bulk of all wind tunnel testing on bridge decks is performed in smooth flows with few exceptions. Since natural winds are turbulent, one might conclude that only turbulent wind tunnel testing is relevant (as is the case for determining wind loads on building frames and cladding). However, results for all anticipated problems typically are conservative when obtained in smooth flow wind tunnels - sometimes extremely conservative. Results so obtained generally can be considered as upper bound conditions, which in many cases may be sufficient for design.

Turbulent flow is typically produced in several ways. Full bridge models have been tested in atmospheric boundary layer wind tunnels where the turbulence is produced by surface roughness elements (905, 908). The scale of the model has been close to the scale of the turbulence so that all aspects of dynamic response and turbulence have been scaled properly. When a large-scale section model is subjected to analogous turbulent flow, the turbulence scale and model scale usually do not match. However, if the turbulence intensity is modeled properly then many aspects of the response may be represented correctly with minimal errors. In some facilities, turbulence has been produced, in otherwise smooth flow, by a system of grids and/or spires (901, 909). Again, the turbulence intensity may be modeled accurately, but the turbulence scale is not likely to be accurate. A unique facility is now being installed where large scale turbulence is being generated by a cascade of moveable vanes (902). One set opens and closes producing longitudinal gusts. The other set produces a vertical gust. Both mechanisms are controlled to produce any desired spectrum of turbulence. This facility represents a state-of-the-art device for the testing of large scale (1/100 to 1/25) section models in properly scaled turbulent flow.

9.2 TORSIONAL DIVERGENCE

The phenomenon of torsional divergence is characterized by a torsional instability, a monotonically increasing rotation to failure at a critical wind speed. The phenomenon is non-oscillatory and can be described in a quasi-steady manner using static aerodynamic data. Like most instabilities, torsional divergence occurs abruptly at the critical wind velocity. It is started by any slight disturbance from the equilibrium position. Natural atmospheric turbulence is a sufficiently large disturbance.

Solution for a critical wind velocity is straightforward given static aerodynamic data, i.e., aerodynamic moments, M, on a given section, about the deck axis, as a function of angle of attack, (901, 911).

Let the moment/rotation characteristics of the bridge deck be given by

$$\{\alpha\} = [C_T] \{M\} \tag{9.2.1}$$

where $\{\alpha\}$ is a vector of deck rotations at intervals along the bridge deck; $\{M\}$ a vector of aerodynamic moments at those intervals; and $[C_T]$ a torsional flexibility matrix where C_{Tij} is the rotation at i due to a unit moment applied at j. The vector of aerodynamic moments is rewritten as

$$\{M\} = 1/2\rho U^2 B^2 \Delta L \frac{\partial C_m}{\partial \alpha} \{\alpha\} \quad (9.2.2)$$

where ρ is the density of the air; U is the mean free stream velocity; B is the deck width; ΔL interval length along the deck (assumed to be constant for simplicity); and $\partial C_m/\partial \alpha$ is the slope of the dimensionless moment curve at $\alpha=0$. It has been assumed that a general dimensionless moment coefficient for all i is given by

$$C_m(\alpha_i) = \frac{\partial C_{m\alpha}}{\partial \alpha} i + C_{mo} \quad (9.2.3)$$

and furthermore that $C_{mo} = 0$, i.e., zero moment at zero angle of attack. For

$$\frac{1}{p} = \frac{1}{2}\rho U^2 B^2 \Delta L \frac{\partial C_m}{\partial \alpha} \quad (9.2.4)$$

Then Equation 9.2.1 can be rewritten to be

$$\{\alpha\} = \frac{1}{p}[C_T]\{\alpha\} \quad (9.2.5)$$

or

$$[[C_T] - p[I]]\{\alpha\} = \{0\} \quad (9.2.6)$$

This is the classical eigenvalue form. Non-zero solutions, which necessarily are non-unique (solutions at torsional divergence) exist if, and only if, the determinant

$$|[C_T] - p[I]| = 0 \tag{9.2.7}$$

The largest eigenvalue corresponds to the lowest critical wind velocity at which torsional divergence will first occur. Denoting this eigenvalue to be p_c, and critical velocity to be U_c then

$$U_c = [2/(\rho p_c B^2 \Delta L \frac{\partial C_m}{\partial \alpha})]^{1/2} \tag{9.2.8}$$

For most bridges, torsional divergence is not a likely form of aerodynamic failure, at least at wind velocities that are likely to occur. For one major bridge, torsional divergence was the governing wind induced instability. However, the critical wind velocity was high enough so as to have a negligible probability of occurrence (905).

It should be noted that the mean moment coefficient needed for this analysis, $C_m(\alpha_i)$, which is a static moment, is not likely to be Reynolds number dependent (if there are sharp edges to the bridge deck which are well defined separation points), and is not likely to differ significantly if obtained in smooth or turbulent flows. The aerodynamic input and the resulting state-of-the-art analysis is expected to be quite reliable.

9.3 VORTEX SHEDDING

It is an interesting phenomenon of fluid mechanics that, for structures that are basically line-like, for flow normal to the structure axis, for certain ranges of Reynolds numbers (based upon a typical section dimensions, D, normal to the structure axis and normal to the direction of flow), there is a strong periodic vortex trail in the wake (901, 904, 909, 911, 912). This phenomenon was first observed by Bénart and Von Karman and commonly bears the name of the latter. This phenomenon was also responsible for the tunes that Aeolus heard when he placed his harp in the wind. Principal characteristics of wake flows behind a bluff body are as follows:

- For very low Reynolds numbers (~ 1) the fluid splits, follows the geometry of the section, and recombines on the leeward side without the formation of cavities in the wake.

- For increasing Reynolds numbers (~ 20) two stable, symmetric vortices are formed behind the section and remain attached to the section.

- For increasing Reynolds numbers (~ 250) vortices of opposite rotation are generated and alternately shed from the section, first of one rotation, then of the other rotation. This periodicity is characterized by the Strouhal number

$$S = \frac{ND}{U} \qquad (9.3.1)$$

  where N is frequency of the periodic vortex shedding into the wake, and U is the mean free stream velocity. As long as the wake width is <u>not</u> strongly Reynolds number dependent (as it most definitely <u>is</u> for circular cylinders) the Strouhal number varies very little over a considerable range of Reynolds number.

- For increasing Reynolds numbers (>1000) the wake becomes more generally turbulent and the strong periodicity becomes less dominant.

- For very high Reynolds numbers ($>10^8$) simple periodicity may be non-existent, but there is increasing spectral evidence that periodicity of a broader-band nature exists.

This vortex shedding phenomenon occurs for bridges as well as any other line-like structure, but because bridges typically are very flexible an otherwise secondary effect becomes significant. Specifically, the wake frequency N will increase linearly with increasing U until a natural frequency of vibration of the bridge is approached. As the periodic wake frequency approaches a natural frequency of the bridge, the bridge is most likely to vibrate at the natural frequency. This oscillation in turn "tunes" the wake periodicity to that frequency. The wake frequency remains locked (referred to as "lock-in" or "lock-on" in the literature) to that natural frequency for a range of velocities. As velocities further increase, the wake frequency will again break away from the natural frequency and will again be shed according to the basic Strouhal number relationship. The range of velocities over which the wake frequency is "locked" to the natural frequency varies directly with the amplitude of vibration. The significance of this "lock-on" effect is that it strengthens the periodicity that would otherwise be weak.

The vortex-induced motion of bridges is an interactive aeroelastic phenomenon rather than a pure forced vibration. The response can be characterized as a nonlinear limit cycle type. The vibrations generally are not catastrophic but may be of an amplitude which is discomforting to pedestrians and passengers, or which may cause fatigue failures (912, 913).

It is an assumed fact that vortex-induced structural motions will exist to some degree. The problem to solve is to estimate the steady-state amplitude of the motions which will result, and to determine whether or not such vibrations are significant. Methods are available to estimate such motions (901, 910, 911). In all cases, time-dependent aerodynamic forces are given in terms of parameters that must be determined experimentally, most often from section model tests. Since section model tests (especially in smooth flow) are performed at Reynolds numbers far lower than those experienced with the full-scale bridge (model Reynolds numbers of 10^4 compared to full scale Reynolds numbers of 10^8, approximately), model results are likely to show periodicity far stronger than that which is expected to occur for the full scale bridge. This is also true because results from section models, rather than full-bridge models, are most often used, and strong coherence along the full bridge span is assumed while in fact it may not exist. It has been shown that the results are not significantly affected if coherence is assumed over the middle half of the span only. However, the assumption of coherence does nothing but add conservatism to the results.

9.4 FLUTTER AND GALLOPING

Flutter is considered in this report to be a vertical, torsional (about the centerline bridge deck axis), or combined vertical/torsional oscillation in the wind due to motion dependent variations in the aerodynamic actions on the bridge deck. Flutter is a true aeroelastic instability where energy is extracted from the mean flow, and the oscillation is not dependent upon pressure fluctuations on the bridge deck due to oncoming turbulence or wake turbulence. Although natural winds are never turbulence-free, nor are wakes behind bridge decks, flutter is an aeroelastic instability which theoretically can exist in absolutely smooth flow (901, 911).

Flutter as it applies to suspension bridge behavior can be divided into two general classifications: Coupled two-degree-of-freedom flutter and single-degree-of-freedom flutter. Coupled vertical and torsional oscillation is similar to what is termed classical flutter in the aircraft industry (911). Streamlined airfoils can exhibit coupled flutter instabilities even though either torsional or vertical flutter motions are stable by themselves. Although the aerodynamics will be different, this type of flutter is possible for suspension bridges. Historically classical flutter has not been a common failure mode.

Single-degree-of-freedom flutter may manifest itself in a vertical or torsional mode but torsional action is by far the more serious. The celebrated failure of the Tacoma Narrows Bridge exhibited two forms of single-degree-of-freedom flutter: initially a non-catastrophic vertical mode; and then, at a higher wind speed, a large-amplitude, catastrophic torsional mode. When airfoils are at, or are near to stall, they are susceptible to a single-degree-of-freedom torsional flutter instability known as "stall flutter" (907). Stall is defined on an airfoil when there is a loss of lift due to the flow separating from the airfoil. Since flow is always separated from typical bridge decks for all angles of attack, it can be expected that single-degree-of-freedom torsional "stall flutter" instability is a real possibility for bridge decks. Historically this has been the case.

A last form of single-degree-of-freedom oscillatory instability which is possible for some forms of bridges is an oscillation in a vertical mode called galloping (911). Galloping occurs at very low reduced frequencies, $B\omega/U$ (B is typical chord or deck width dimension, ω is the frequency of vibration, and U is the mean free stream velocity). Galloping is phenomenologically different from other forms of single-degree-of-freedom instability in a vertical mode in that it tends to be associated with large amplitudes. Because the reduced frequency is low, aerodynamic pressure distributions over the bridge deck vary almost in unison with variations in the bridge motion. Therefore, mean or average static aerodynamic data (lift and drag coefficients) may be used to assess the susceptibility of a particular bridge deck to the galloping phenomena. Galloping typically is not an instability mode for highway bridges but may be a possible vibration mode for a pipeline bridge, particularly for a bundle of pipes which may not form a circular cross-section. Forms of galloping are a common instability mode for transmission lines that are bundled ("along-wind" galloping), that have a cable support, or that have non-circular cross-sections. Very large steady state oscillations are possible.

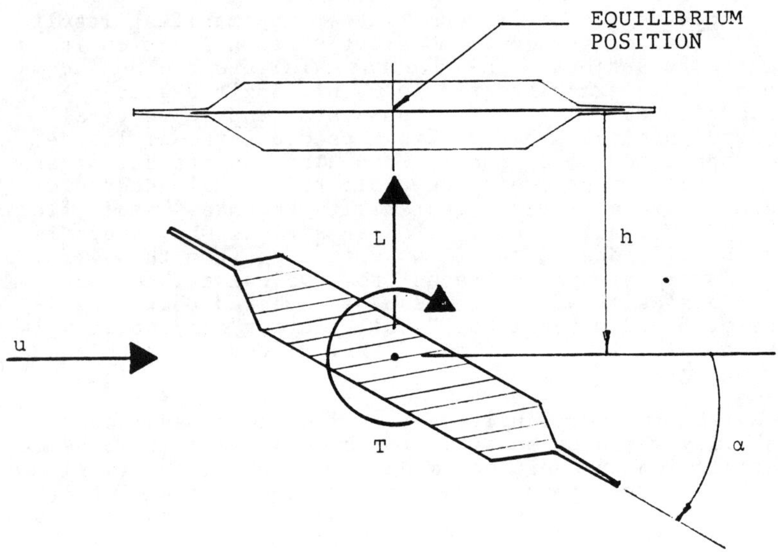

Figure 9.4.1 Definition of Positive Actions and Positive Coordinate Directions

Various possible modes of oscillatory instability exist caused by one structure being in the wake of another. These forms, associated with shear flows, cannot be generalized readily. However, the designer or analyst should be aware of their existence when one flexible structure is in close proximity to another. Experimental methods most probably must be employed to investigate these problems further.

GENERAL PROBLEM FORMULATION

In the usual manner, it will be assumed that the most critical wind is horizontal and perpendicular to the axis of the roadway. Furthermore it is assumed, at a given longitudinal location, x, that the only significant degrees of freedom are a vertical translation, h(x,t), and a rotation about the longitudinal bridge deck axis, x, (x,t). These degrees of freedom and the positive directions of aerodynamic moment and lift are shown in Figure 9.4.1.

In all forms, flutter usually occurs in a single natural mode of vibration, or a coupled pair of a single vertical mode and a single torsional mode. Therefore, it is reasonable to write

$$h(x,t) = h(x)p(t) \tag{9.4.1}$$

$$\alpha(x,t), = \alpha(x)q(t) \tag{9.4.2}$$

Suspension bridge decks typically are symmetric (or nearly symmetric) about the longitudinal bridge axis. Therefore, translational and rotational <u>mechanical</u> forces are expected to be uncoupled. The equations of motion take the form

$$M_1[\ddot{p} + 2\xi_h\omega_h\dot{p} + \omega_h^2 p] = L_1 \tag{9.4.3}$$

$$I_1[\ddot{q} + 2\xi_\alpha\omega_\alpha\dot{q} + \omega_\alpha^2 q] = T_1 \tag{9.4.4}$$

where the following definitions are made:

ξ_h, ξ_α damping ratios;

ω_h, ω_α natural circular frequencies of vibration;

$$M_1 = \int_o^L m(x)h^2(x)dx \;;$$
$$I_1 = \int_o^L I(x)\alpha^2(x)dx \;;$$

and L_1 and T_1 are modal aerodynamic actions, a lift and a torque respectively.

Although it is likely that the aerodynamic actions will be highly nonlinear, for small motions they may be approximated well as being linear. It should be noted that, while the spectacular motions of the Tacoma Narrows bridge first come to mind, it is the minute oscillations at incipient flutter which are technically the most interesting. Therefore, to determine the velocity at which incipient flutter occurs it is sufficient to study the stability of small oscillations for which a linear approximation is accurate.

It has been shown that the aerodynamic actions can be approximated well, for loads on suspension bridges, as a linear combination of $\dot{h}$, α, and $\dot{\alpha}$ at any point x on the bridge (910, 911). The aerodynamic inertial terms may be significant relative to the mechanical inertial terms for aircraft airfoils, but they typically are not for suspension bridges.

The modal aerodynamic actions can be presented in the following form:

$$L_1 = H_1\dot{p} + H_2\dot{q} + H_3q \qquad (9.4.5)$$

$$M_1 = A_1\dot{p} + A_2\dot{q} + A_3q \qquad (9.4.6)$$

or in dimensionless terms more consistent with classical aerodynamics

$$L_1 = 1/2\rho U^2(2B)[KC_{11}H_1^*\dot{p}/U + KC_{12}H_2^*B\dot{q}/U + K^2C_{12}H_3^*q] \quad (9.4.7)$$

$$M_1 = 1/2\rho U^2(2B^2)[KC_{12}A_1^*\dot{p}/U + KC_{22}A_2^*B\dot{q}/U + K^2C_{22}A_3^*q] \quad (9.4.8)$$

where

ρ		density of air
B		deck width
K	=	$B\omega/U$
ω		circular frequency of oscillation
C_{11}	=	$\int_0^L h^2(x)dx$
C_{12}	=	$\int_0^L h(x)\alpha(x)dx$
C_{22}	=	$\int_0^L \alpha^2(x)dx$

The coefficients A_1^*, A_2^*, A_3^*, H_1^*, H_2^*, and H_3^* are motional derivatives which must be obtained experimentally. In Japan these have been obtained using forced vibration tests of a section model. Typically in the United States these motional derivatives have been obtained using simple free vibration tests with sectional models. Only for airfoils at angles of attack less than stall can and have these derivatives be obtained theoretically from fundamental principles (903, 905, 907, 908, 910, 912).

As described previously, often suspension bridge flutter motion is dominated by single-degree-of-freedom motion in the torsional or vertical mode. In either case, the problem formulation follows obviously from the formulation for the general two-degree-of-freedom problem. Specifically, the two single-degree-of-freedom equations of motion become

$$M_1[\ddot{p} + 2\xi_h\omega_h\dot{p} + \omega_h^2 p] = 1/2\rho U^2(2B)[KC_{11}H_1^*\dot{p}/U] \quad (9.4.9)$$

and

$$I_1[\ddot{q} + 2\xi_\alpha\omega_\alpha\dot{q} + \omega_\alpha^2 q] = 1/2\rho U^2(2B^2)[KC_{22}A_2^*B\dot{q}/U + K^2C_{22}A_3^*q] \quad (9.4.10)$$

The first of the above two equations holds true for single-degree-of-freedom vertical motion, or for problems of galloping type. In the general case, the aerodynamic lift

$$L_1 = 1/2\rho U^2(2B)[KC_{11} H_1^* \dot{p}/U] \qquad (9.4.11)$$

is dependent upon the reduced frequency, K. Specifically (KH_1^*) varies as a function of K. As discussed earlier, galloping is a form of single-degree-of-freedom vertical motion that occurs at very low values of K. Furthermore, for such cases, the term (KH_1^*) is independent of K and is traditionally given in terms of lift and drag coefficients for the bridge deck. Again, these terms are obtainable from static aerodynamic tests alone. Specifically

$$(KH^*) = -(\frac{dC}{d\alpha}L + C_D) \qquad (9.4.12)$$

in which the Den Hartog sum of lift slope and drag coefficient appears. Note that the lift and drag coefficients for galloping typically are given in terms of a vertical dimension rather than an along wind dimension, B. Care should, therefore, be used in transferring from one form to the other.

GENERAL PROBLEM SOLUTION

The general solution method to Equations 9.4.7 and 9.4.8 is explained in detail in References 910 and 911, and is summarized in the following paragraphs. The solution method to the single-degree-of-freedom problem is theoretically identical, but is practically far simpler. It will be discussed briefly at the end of the section.

Because motions, at incipient flutter, are small and are dominated heavily by the mechanical system which has been assumed to be linear, it is reasonable to assume that p and q have solutions in the form $e^{i\omega t}$ where ω is in general complex, $\omega = \omega_1 + i\omega_2$. The discplacements

$$p(t) = p_0 e^{i\omega t} \qquad (9.4.13)$$

$$q(t) = q_0 e^{i\omega t} \qquad (9.4.14)$$

are substituted into Equations 9.4.7 and 9.4.8. A reduced frequency of oscillation K is assumed, and the coefficients A_i^*, H_i^* (i=1, 2, 3) are determined for a first approximation. The determinant of coefficients of the amplitudes, p_0 and q_0, is set equal to zero as the basic stability criterion. The complex frequency ω is the solution sought. If the imaginary part ω_2 is other than zero then the oscillation is a divergent or a decaying oscillation, and is not representative of incipient flutter. A systematic, iterative approach is then used, each time assuming a new value of K, to determine a solution for which $\omega=\omega_1$ and $\omega_2=0$. The critical flutter velocity U_c is then determined to be

$$U_c = B\omega_1/K_c \tag{9.4.15}$$

where K_c is the critical reduced frequency at which ω_2 is found to equal zero.

Solution for single-degree-of-freedom vertical motion flutter is similar and far more straightforward. For

$$M_1[\ddot{p} + 2\xi_h\omega_h\dot{p} + \omega_h^2 p] = 1/2\rho U^2(2B)[KC_{11}H_1^*\dot{p}/U] \tag{9.4.16}$$

the stability is determined completely by the sign of the viscous damping term

$$(M_1(2\xi_h\omega_h) - 1/2\rho U^2(2B)KC_{11}H_1^*/U) \tag{9.4.17}$$

If this term is positive, then any perturbation will decay, and the bridge will be stable. If this term is negative then any perturbation will diverge, and the bridge will be unstable. Obviously, therefore, incipient flutter will occur for a value of K_c at which this damping term is zero. Again

$$U_c = B\omega_h/K_c \tag{9.4.18}$$

where ω_h is the natural circular frequency of the vertical motion.

Single-degree-of-freedom torsional flutter is similarly dominated by the coefficient of the $\dot{q}$ term, including mechanical and aerodynamic parts. The only difference is that the frequency of oscillation is also dependent upon the wind velocity U because there is a mechanical as well as an aerodynamic coefficient to q. The frequency of vibration is not likely to be substantially different from ω_α, in which case the critical flutter velocity is dominated by the magnitude and sign of A_2^*.

It can be concluded that the state-of-the-art solution methods to suspension bridge flutter problem are highly refined. Solution errors are limited by the accuracy of the experimental coefficients, H_i^* and A_i^* (i=1,2,3) and the limitations imposed by the assumption of linearity. The latter is an accurate approximation for incipient flutter, and it is the detection of incipient flutter which is of utmost importance. Prediction of large amplitude, highly nonlinear, catastrophic bridge deck motions is not readily possible, but then again, it is of academic interest only. The experimentally obtained motion derivatives are most likely obtained with acceptable accuracy (±10%) but care should be exercised in their use. Because most bridge deck sections do have sharp edges, one is inclined to think that aerodynamic results should be Reynolds number independent. However, as discussed previously, these aerodynamic coefficients are dependent upon vortex formation in the wake and flow through small passages. Such aerodynamic conditions are Reynolds number dependent. Therefore, care should be taken to perform experimental tests at Reynolds numbers as close to full scale values as possible.

9.5 BUFFETING

Buffeting typically is defined as the randomly forced vibration of a structure due to velocity fluctuations in the oncoming wind. Again, typically, it is a pure forced vibration where the forcing function is totally independent of the structure motion. This definition of buffeting applies as well to suspension bridges with one exception: equivalent modal damping and natural frequencies including aerodynamic damping and aerodynamic stiffness are used instead of pure mechanical damping and mechanical stiffness. These aerodynamic terms are the same as those defined in the previous section of flutter, i.e., H_i^*, A_i^*, (i=1,2,3). The following assumptions are made:

- It is usual to assume that the structure behaves in a linear manner and that any response can be composed of a linear combination of normal modal responses.

- It is usual to assume that the velocity fluctuations in the oncoming wind can be characterized as a stationary and ergodic stochastic process. When a suspension bridge is subjected to such a wind it is usual to determine statistics of the steady-state response of the bridge to a given wind velocity (rather than to determine a critical wind at which an instability occurs).

- It is usual to assume that velocity fluctuations in the oncoming flow have periods which are long compared to the natural periods of the bridge so that fluctuating forces and moments on the bridge deck can be computed using the fluctuating velocity and steady-state lift, drag, and moment coefficients (C_L, C_D, and C_M) on a typical bridge deck.

Following is a description of the solution procedure (911): A given wind must first be defined which has a mean velocity, U, and fluctuating components u(t), v(t), and w(t). For each of the fluctuating components a power spetral density function must be defined. For each mode of vibration of the bridge deck (and in this section it is assumed that there is no aerodynamic coupling between natural modes of vibration) generalized actions are determined in terms of the bridge deck geometry, the steady-state aerodynamic coefficients (C_L, C_D, and C_M), the mode shape, and the time dependent wind velocity. In order to compute the generalized actions it is necessary to make two additional assumptions. First, the correlation between two fluctuating wind velocities is assumed to decay exponentially from unity to zero as a function of the separation distance between those two points. Reference 911 suggests reasonable decay rates for the correlations. Second, the fluctuating velocities are assumed to be small when compared to the mean wind velocity. It, therefore, can be concluded that the power spectral density of the fluctuating generalized action is a linear combination of the power spectral densities of the three fluctuating velocity components, u(t), v(t), and w(t).

It is then required to compute equivalent natural frequencies and damping ratios taking into consideration aerodynamic stiffness and aerodynamic damping. It is assumed that the motional derivatives, i.e., H_i^*, A_i^* (i=1,2,3), are constant and a function of the equivalent natural frequency and the mean velocity, U. It is furthermore assumed that motional derivatives obtained in smooth flow yield conservative results (901, 903, 905, 906, 910, 911). When the variable vane turbulence generator (similar to that in Reference 902) is installed in the George S. Vincent Wind Tunnel of the Federal Highway Administration, motional derivatives in correctly scaled turbulence will be obtainable.

For a pure torsional mode of vibration the equivalent frequency of vibration and damping ratios are of the form

$$\omega_{ei}^2 = \omega_i^2 - c_1 A_3^* (K) \qquad (9.5.1)$$

and

$$\xi_{ei} = \xi_i - c_2 A_2^*(K) \qquad (9.5.2)$$

where ω_i and ξ_i are the natural frequency and damping ratio for the ith mode of vibration, and c_1 and c_2 are constants. These two constants are to be computed from the bridge geometry, mode shape, frequency of vibration and mean wind velocity.

At this point the problem is formulated like the formulation for any other single-degree-of-freedom oscillator subjected to a stationary and ergodic stochastic forcing function. Solution methods are well documented in many textbooks and will not be duplicated here. The results of such an analysis typically are a power spectral density function of the modal response and, consequently, its variance.

At the present time, the state-of-the-art solution method of a suspension bridge being buffeted by winds is as outlined above. The formulation can be readily extended to multi-degree-of-freedom systems that are aerodynamically coupled. The solution of such problems, while possible in closed form, is most readily obtained numerically on a computer. Alternative solutions may also be obtained by observation of a dynamically, scaled, aeroelastic, full-bridge model in a wind tunnel under the action of properly simulated turbulent wind.

9.6 SUMMARY

In this chapter have been discussed four aeroelastic phenomena, and the corresponding state-of-the-art solution procedures, that must be considered in the design of long, flexible bridges. Not all phenomena are equally likely to occur, and certainly not all phenomena lead to catastrophic failures.

Torsional divergence is a phenomenon that is very unlikely to occur for bridge sections of typical mass and typical stiffness. Torsional divergence is most likely to occur for very lightweight aircraft type structures.

Vortex shedding induced vibrations commonly occur for pipeline suspension bridges, and components (hangers and cables) on other bridges. Failures from vortex induced vibrations rarely are catastrophic, but they may produce discomfort or they may lead to fatigue failures. Since all natural winds are turbulent, and since turbulence inhibits the formation of strong vortices, vortex shedding analyses assuming smooth and uniform flow are generally very conservative.

Classical flutter and galloping are phenomena most feared because failures are often catastrophic. Galloping is most common for ice covered power lines or bundled power lines. Galloping is an infrequent mechanism of bridge failure. Single-degree-of-freedom torsional flutter is the most likely form of bridge flutter and the one form most likely to lead to catastrophic failure. State-of-the-art procedures for the solution of single-degree-of-freedom torsional flutter problems are conservative if motional derivatives are obtained in smooth flows. Use of turbulent flow wind tunnels to obtain motional derivatives will reduce this conservatism. Single-degree-of-freedom vertical translational flutter is likely to occur, but rarely leads to catastrophic failure. Classical, coupled flutter occurs infrequently for bridges.

REFERENCES

901 Milne, M., et al, Bridge Aerodynamics, Conference Proceedings, Institute of Civil Engineers, Thomas Telford Limited, London, 1981.

902 Bienkiewicz, B., Cermak, J. E., and Peterka, J. A., "A New Technique of Modeling Atmospheric Turbulence for Wind-Tunnel Tests of Bridge Models", NSF Grant Number CME8013977, Fourth U. S. National Conference on Wind Engineering Research, National Science Foundation Wind Engineering Research Council and the University of Washington, Seattle, Washington, July, 1981.

903 Bosch, H., "Aerodynamic Investigations of the Luling, Louisiana Cable-Stayed Bridge", Report No. FHWA-RD-77-161, Federal Highway Administration, Washington, DC, July 1978.

904 Chi, M., "Response of Bridge Structural Members Under Wind-Induced Vibrations", Report No. FHWA-RD-78-25, Federal Highway Administration, Washington, DC, June 1976.

905 Davenport, A. G. et al, "A study of Wind Action on a Suspension Bridge During Erection and on Completion: The Narrows Bridge, Halifax, Nova Scotia, Canada", BLWT-3-69, Boundary Layer Wind Tunnel Laboratory, The University of Western Ontario, London, Canada, May 1969.

906 Davenport, A. G., "Comparison of Model and Full Scale Tests on Bridges", Wind Tunel Modeling for Civil Engineering Applications, Proceedings of the International Workshop on Wind Tunnel Modeling Criteria and Techniques in Civil Engineering Applications, Gaithersburg, Maryland, USA, April, 1982, Cambridge University Press.

907 Raggett, J. D., "Linear Theory of Stall Flutter", AIAA Journal, Vol. II, No. 5, May 1973, pp 733-735.

908 Scanlan, R. H., "Aeroelastic Modeling of Bridges", Wind Tunnel Modeling for Civil Engineering Applications, Proceedings of the International Workshop on Wind Tunnel Modeling Criteria and Techniques in Civil Engineering Applications, Gaithersburg, Maryland, USA, April, 1982, Cambridge University Press.

909 Scanlan, R. H., "Recent Methods in the Application of Test Results to the Wind Design of Long, Suspended-Span Bridges", Report No. FHWA-RD-75-115, Federal Highway Administration, Washington, DC, October 1975.

910 Scanlan R. H., "State-of-the-Art Methods for Calculating Flutter, Vortex-Induced, and Buffeting Response of Bridge Structures", Report No. FHWA/RD-80-050, Federal Highway Administration, Washington, DC, April 1981.

911 Simiu, E., and Scanlan, R. H., Wind Effects on Structures, John Wiley & Sons, New York, 1986, Second Edition

912 Wardlaw, R. L., and Goettler, L. L., "A Wind Tunnel Study of Modifications to Improve the Aerodynamic Stability of the Longs Creek Bridge", LTR-LA-8, National Aeronautical Establishment, Ottawa, Canada, August 1968.

913 Wardlaw, R. L., and Ponder, C. A. "Wind Tunnel Investigations of the Aerodynamic Stability of bridges", LTR-LA-47, National Aeronautical Establishment, Ottawa, Canada, February 1970.